Advances in Artificial Intelligence in Education

Volume 2

This book series focuses on using artificial intelligence (AI) in educational contexts, providing the venue to showcase the latest developments in this rapidly moving field. From personalised instruction to adaptive assessments and immersive learning experiences, AI tools and technologies can revolutionise education and empower educators, students, and decision makers.

A wide range of topics surrounding the use of AI in learning and teaching are explored, covering global perspectives from different communities and disciplines. The book series approaches the use of AI in education with caution and addresses ethical considerations such as data privacy, algorithmic bias, and transparency. By examining AI socially responsible and inclusively, AI's full potential shall be unlocked and create a future of education that is personalised, engaging, and equitable for all learners.

The titles in the series cover the use of AI within K-12 through tertiary settings and in other domains such as professional, lifelong, and work-integrated learning. The series accepts monographs and edited volumes that fit the scope mentioned above and covers various topics, including (but not limited to) tutoring systems, adaptive learning, feedback provision, learner modelling, assessment analytics, as well as teaching and learning support. The series also includes titles covering related issues of ethical and responsible use of AI, data privacy, bias and transparency, educational design, institutional adoption, and inclusion.

Please contact Professor Dirk Ifenthaler at dirk@ifenthaler.info if you wish to discuss a book proposal for the series.

Xue Zhou • Hosam Al-Samarraie
Editors

Institutional guide to using AI for research

Springer

Editors
Xue Zhou
University of Leicester
Leicester, UK

Hosam Al-Samarraie
University of Leeds
Leeds, UK

ISSN 3004-9903 ISSN 3004-9911 (electronic)
Advances in Artificial Intelligence in Education
ISBN 978-3-031-94808-4 ISBN 978-3-031-94809-1 (eBook)
https://doi.org/10.1007/978-3-031-94809-1

This Springer imprint is published by the registered company Springer Nature Switzerland AG
The registered company address is: Gewerbestrasse 11, 6330 Cham, Switzerland

Preface

In the rapidly evolving landscape of higher education, artificial intelligence (AI) has emerged as a transformative force. This has significantly reshaped the way we approach academic research. As students embark on their academic journey, it is common for them to find themselves at the forefront of this technological revolution, attempting to take advantage of AI to enhance their research capabilities and contribute to the advancement of knowledge in their chosen fields.

This book, *Institutional Guide to Using AI for Research*, is designed to provide a general perspective on how higher education institutions (HEIs) could navigate the exciting yet complex world of AI in research. Our aim is to equip HEIs with the knowledge, skills, and ethical framework necessary to leverage AI tools (e.g., generative AI) effectively and responsibly in their academic pursuits. The integration of AI into academic research is not merely a trend; it represents a fundamental shift in how we process information, analyse data, and generate insights. From natural language processing algorithms that can sift through vast amounts of literature in seconds to machine learning models that can identify patterns in complex datasets, generative AI is expanding the boundaries of what is possible in research across all disciplines. Yet, this also brings a number of challenges and concerns among both educators and students, with great power comes great responsibility. As we delve into the applications of AI in research, we must also grapple with the ethical implications and potential pitfalls of relying too heavily on these technologies. This book seeks to strike a balance between embracing the potential of AI and maintaining the critical thinking and human insight that are at the core of academic inquiry. Throughout this guide, we will explore the multifaceted role of AI in academic research, providing you with both theoretical understanding and practical skills.

The book is structured into four key sections, each addressing a crucial aspect of AI in research:

Part I: The Role of AI in Advancing Research in Higher Education

In this section, we lay the foundation for understanding AI's place in the academic landscape. Chapter 1 provides an introduction to AI in research, outlining its evolution and current capabilities. Chapter 2 delves into the perception of generative AI in higher education, exploring both the enthusiasm and scepticism surrounding these technologies. Chapter 3 offers an overview of the AI tools and technologies available for academic research, giving you a practical understanding of the resources at your disposal.

Part II: Practical Applications and Excellence in AI-Driven Research

This section focuses on the hands-on application of AI in various stages of the research process. Chapter 4 explores how AI can possibly revolutionise literature reviews, helping you navigate the ever-growing sea of academic publications. Chapter 5 explores AI's role in data analysis, showcasing how machine learning algorithms can uncover patterns and insights that might elude human researchers. Chapter 6 addresses the critical task of interpreting research outcomes with AI, emphasising the importance of human judgement in contextualising AI-generated results. Finally, Chapter 7 covers AI-assisted writing and editing, a topic of particular relevance in the age of large language models and automated writing tools.

Part III: Ethical Considerations in AI-Enabled Research

As we harness the power of AI, we must also confront the ethical challenges it presents. Chapter 8 explores the broader ethical implications and governance of AI in research, encouraging you to think critically about the impact of these technologies on academic integrity and societal outcomes. Chapter 9 focuses specifically on data integrity and AI ethics, addressing crucial issues such as bias in AI models, privacy concerns, and the responsible use of data in research.

Part IV: Policy Perspectives and Future Directions in AI and Higher Education

The final section of the book looks towards the future, examining the evolving landscape of AI in academia. Chapter 10 discusses AI policies in academic institutions, providing insight into how universities and research organisations are adapting to the AI revolution. Chapter 11 explores the horizon of AI in academic research, offering a glimpse into emerging technologies and potential future developments that may shape your research careers.

Readers of this book are encouraged to approach the material with both enthusiasm and critical thinking. The knowledge introduced in this book aims to provide basic means for students to ethically and responsibly experiment with AI while conducting their research. As you progress through this book, you'll develop not only technical skills but also a deeper understanding of the research process itself. You'll learn how to formulate better research questions, design more robust studies, and communicate your findings more effectively—all with the assistance of AI tools. We have designed this guide to be accessible to students across various disciplines, recognising that AI's impact on research is not limited to any single field. Whether you're studying computer science, humanities, social sciences, or natural sciences, you'll find valuable insights and practical advice tailored to your needs. Each chapter includes examples and case studies to help you apply the concepts to your own research projects. We've also included discussions on the limitations and potential biases of AI systems, encouraging you to approach these tools with a discerning eye.

Remember that AI tools, no matter how advanced, are just that—tools. They are meant to augment your abilities, not replace your creativity, critical thinking, or ethical judgement. As you learn to leverage these powerful technologies, always strive to maintain the human touch that is essential to meaningful research and discovery.

Leicester, UK Xue Zhou
Leeds, UK Hosam Al-Samarraie

Contents

Contributors

Tasnim Ahmed Liverpool John Moores University, Liverpool, UK

Hosam Al-Samarraie University of Leeds, Leeds, UK

Lucy Caton The University of Greater Manchester, Bolton, UK

Khaldoon Dhou Texas A&M University Central Texas, Killeen, TX, USA

Joana Eva Dodoo University of Cape Coast, Cape Coast, Ghana

Olatunde Durowoju Liverpool John Moores University, Liverpool, UK

Pedro Elston Queen Mary University of London, London, UK

Aniekan Essien University of Bristol, Bristol, UK

Lei Fang Queen Mary University of London, London, UK

Nick Fisher Queen Mary University of London, London, UK

Paula Funnell Queen Mary University of London, London, UK

Griffiths Twum Appeaning University of Cape Coast, Cape Coast, Ghana

Yujeong Hwang The University of Greater Manchester, Bolton, UK

Dániel Kuttor Faculty of Economics, University of Miskolc, Miskolc, Hungary

Cassandra Lewis Queen Mary University of London, London, UK

Leyla Orudzheva Texas A&M University Central Texas, Killeen, TX, USA

Zsolt Pál Faculty of Economics University of Miskolc, Miskolc, Hungary

Lilian Schofield Queen Mary University of London, London, UK

Zhenyu Sun University of Chinese Academy of Sciences, Beijing, China

Attila Szendi Library, Archives, Museum, University of Miskolc, Miskolc, Hungary

Ruixue Yuan Beijing University of Chemical Technology, Beijing, China

Xue Zhou University of Leicester, Leicester, UK

Part I
The Role of AI in Advancing Research in Higher Education

Chapter 1
Introduction to AI in Research

Xue Zhou and Hosam Al-Samarraie

Abstract This chapter introduces the transformative role of Artificial Intelligence (AI) in reshaping research methodologies across various academic fields. We delve into fundamental AI concepts, its evolution, its growing significance in academic research, and its impact on enhancing research methodologies. As AI technologies become increasingly integral to the research process—from literature review and data collection to analysis and hypothesis generation—students, academics, and professionals must grasp the profound implications and opportunities AI presents. This chapter also addresses the critical challenges of data privacy, algorithmic bias, and the interpretability of AI-generated results, which are paramount in ensuring responsible and ethical research. Through a comprehensive overview, this chapter equips students with the necessary knowledge and critical thinking skills to navigate the complex landscape of AI in research. We explore the impact of AI on developing research skills, the ethical considerations in its application, and future trends in AI-driven research. By the end of this chapter, students will better understand AI's role in modern research, enabling them to critically evaluate the benefits and limitations of AI tools in their respective fields.

1 Introduction of the Chapter

The recent advancement in Artificial Intelligence (AI) has led to a new era of scientific practices that is fundamentally transforming how we approach discovery, analysis, and innovation across all fields of study. As we stand at the point of this AI

X. Zhou (✉)
University of Leicester, Leicester, UK
e-mail: xue.zhou@le.ac.uk

H. Al-Samarraie (✉)
University of Leeds, Leeds, UK
e-mail: h.alsamarraie@leeds.ac.uk

X. Zhou, H. Al-Samarraie (eds.), *Institutional guide to using AI for research*,
Advances in Artificial Intelligence in Education 2,
https://doi.org/10.1007/978-3-031-94809-1_1

revolution in research, students, academics, and professionals alike must understand the profound implications and opportunities that AI presents in the area of scientific inquiry (Golan et al., 2023). Understanding AI in the context of research is no longer a luxury but a necessity. AI technologies are rapidly becoming integral to every stage of the research process, from literature review and data collection to analysis and hypothesis generation (Khalifa & Albadawy, 2024). These tools are not merely automating routine tasks; they are augmenting human intelligence, enabling researchers to tackle more complex problems, uncover hidden patterns in vast datasets, and accelerate the pace of discovery in ways previously unimaginable (Perkins & Roe, 2024a).

Since AI is multifaceted and ever-expanding, emerging issues of data privacy, algorithmic bias, and the interpretability of AI-generated results are at the forefront of discussions in the scientific community (Hacker et al., 2024). Understanding these challenges is crucial for conducting responsible, ethical research in the AI age. Thus, the purpose of this chapter is to provide a comprehensive overview of AI in research, equipping students with the knowledge and critical thinking skills necessary to navigate this new landscape. Throughout this chapter, we will explore:

- The fundamental concepts of AI and how they apply to assist research.
- The impact of AI on research skills development, from data analysis to critical thinking.
- The ethical considerations and challenges in integrating AI into research processes.
- Future trends and potential developments in AI-driven research.

By the end of this chapter, students can expect to have gained an understanding of the role AI plays in modern research. They will be able to critically evaluate the potential benefits and limitations of AI tools in their own fields of study. Moreover, they will develop an appreciation for the interdisciplinary nature of AI-driven research and the importance of combining domain expertise with AI literacy.

2 AI in Research

AI generally refers to computer-controlled machines that perform tasks similar to those carried out by humans. In research, it could refer to the tools, skills, and practices that are commonly used by students to understand aspects related to their research. Generative AI (GenAI) is the latest form of AI that generates contents in response to users in the form of text, images, or audio using a large training data. ChatGPT and Dall-E are an example of GenAI that serves. These tools are extensively used in the higher education sector to support various stages of research development, but their use, value and challenges for the research community keeps on changing. This is because AI keeps on evolving as a result of increased in individuals' digital footprint and usage. This increase has resulted in reforming the way AI learn and respond to diverse needs of individuals.

This chapter shades the light on the potential of GenAI in research workflow. Figure 1.1 break down the stages of where GenAI has been reported in the literature to be used in research practices. From the Figure, it can be seen that GenAI play a role in supporting researchers at an early stage of their research. In the proposal stage, students have been widely considered to take advantage of GenAI to ensure their ideas and narratives are presented in academic manners. For example, GenAI has been reported to offer the means for students to create a structure for their research proposal and help draft parts or refine or edit language of it (Andersen et al., 2024). Ideation, synthesis of previous studies, reflection, and writing were the main topics discussed by previous studies at the proposal writing stage with the support of GenAI. A significant aspect of this viewpoint was the potential of generative tools to enhance students' idea generation and writing skills by providing suggestions and offering feedback on grammar, sentence structure, and word choice (Al-Samarraie et al., 2024). It is worth noting that the inclination to use GenAI tools at the proposal stage might stem from the ability of these tools to drive motivation, flexibility, and competency among users of specific learning needs.

Our review of the literature, seeding activities (university wide GenAI brainstorming sessions), and conversations with key educational organizations in the UK have led to the understanding that GenAI can be widely used to:

1. Identify the need for knowledge and clarify ethical considerations.
2. Collect relevant information and generate data following appropriate methodologies.
3. Evaluate information/data and reflect on the processes used.
4. Organise information and manage research stages.
5. Analyse information/data and synthesize new knowledge to produce coherent understandings.
6. Communicate the processes, understandings, and applications of the research, and respond to feedback while being mindful of ethical, cultural, and social issues.

Literature writing, data analysis, results interpretation, and reporting have been recognised as key research practices where students relay on GenAI tools to progress in their research. These applications of GenAI vary significantly across disciplines, which is largely linked to the choice of methodology adopted in each field to seek new knowledge. For example, in the humanities, students may employ GenAI to refine language and structure in their academic writing, while in the sciences, they may rely on AI for complex data analysis and simulation modelling. In both cases, the use of GenAI has been found to help—to some level—streamline research processes by offering quick access to relevant information, generating data insights, and ensuring clarity and coherence of findings.

Yet, to mitigate the potential misuse of GenAI tools in forming irrelevant or inaccurate research directions, many doctoral colleges across the sector have acknowledged the lack of unified policies to clearly define what GenAI tools are acceptable for use in research writing and how they should be implemented. In addition, there is a significant gap in available training resources to guide students and staff in effectively and ethically using these technologies. Such lack has been perceived by

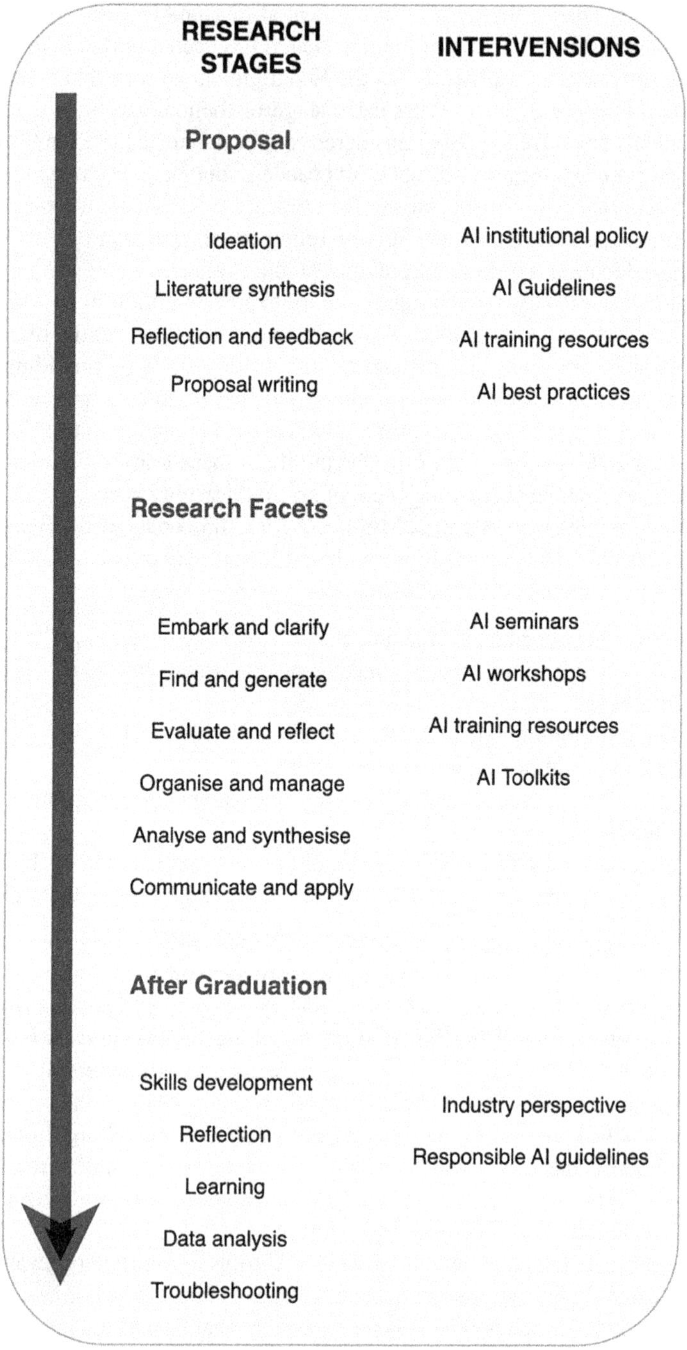

Fig. 1.1 GenAI in research and needed interventions

academics as a concern that potentially can lead to inconsistent practices. In addition, the over reliance on GenAI at an early stage of research can pose several novelty and integrity related concerns for the sector and potential supervisors. On its core the lack of developing skills essential for students to fully comprehend how specific research steps can contribute to their research journey. Therefore, institutions are increasingly recognising the need for comprehensive policies and tailored training programs to ensure responsible and beneficial integration of GenAI tools in academic research.

3 Bibliometric Analysis of GenAI in Research

To explore the evolving nature of GenAI, we conducted a bibliometric analysis of studies focusing on its use in research. Using R, we analysed and visualised trends, sentiments, and topics from research articles retrieved from the Scopus database using keywords that reflect GenAI and research practices. We began by creating a .bib file containing 2520 studies, including details such as authors, titles, keywords, abstracts, and publication years. This .bib file was then imported into R for topic modelling to identify key terms in the literature. Several preprocessing steps were carried out in R to clean and normalise the text data. From there, we used a Latent Dirichlet Allocation model to extract a document-topic matrix, helping us understand how topics were distributed across the documents. We visualised the top terms for each topic. We also analysed the sentiments expressed in the abstracts of these studies, which were extracted and visualised. Figure 1.2 illustrates the volume of

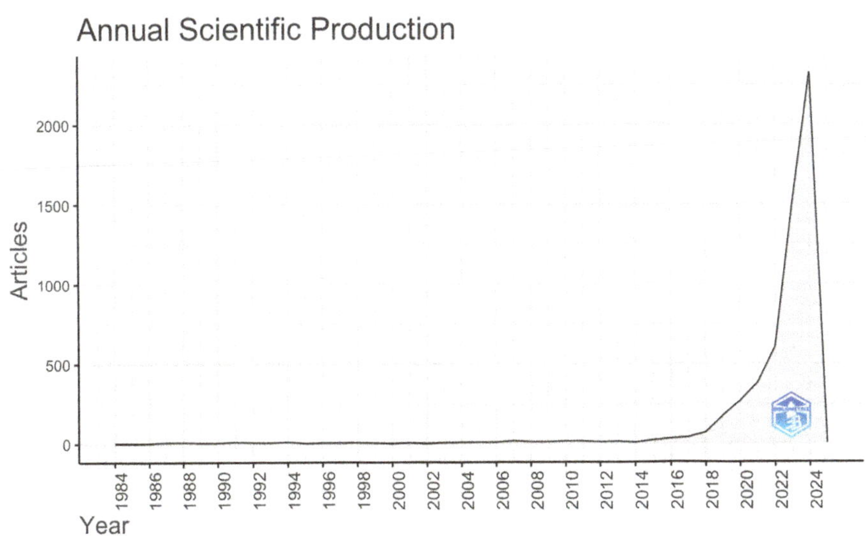

Fig. 1.2 GenAI publications in research by year

research published on the use of GenAI in research, showing a significant increase from 2020 onward. This surge is likely due to the demand for GenAI-powered online solutions during and after the Covid-19 pandemic.

The abstracts of the 2520 studies were extracted and processed for sentiment analysis. The results (see Fig. 1.3) showed that sentiment around the application of GenAI in research was largely positive. Positive emotions, such as trust and joy, were found to drive students' optimism and confidence in GenAI's potential to enhance their research practices. Anticipation and excitement were also common, indicating enthusiasm about future GenAI-driven tools. However, negative sentiments like fear and sadness were also notable, likely reflecting concerns about ethical challenges, potential misuse, and impacts on traditional research methods. Meanwhile, the lower levels of anger and disgust may indicate frustrations with GenAI's limitations, such as its accuracy in locating and comparing sources of information. These findings highlight the need for a balanced approach, recognising both the opportunities and the ethical or practical concerns GenAI raises.

The analysis of the top positive terms related to the use of GenAI in research—as shown in Fig. 1.4—revealed a strong focus on aspects that enhance efficiency and cognitive processes. Terms such as "simplicity," "accessibility," and "speed" underscore the value students place on GenAI's ability to streamline tasks and make

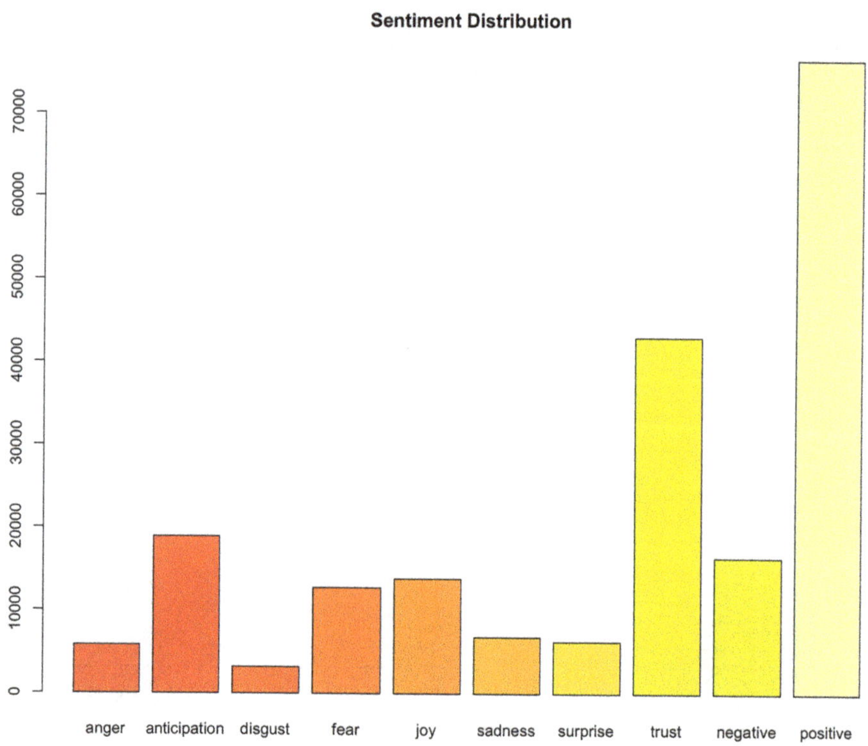

Fig. 1.3 Sentiment results of GenAI use in research

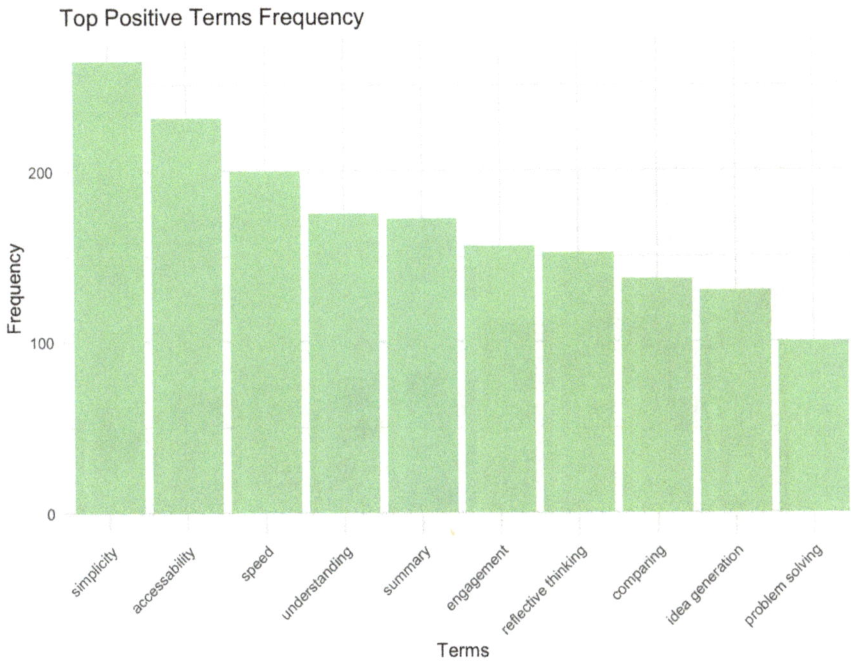

Fig. 1.4 Top positive terms on the use of GenAI in research

research tools more readily available and faster. The emphasis on "understanding," "reflective thinking," and "engagement" suggests that GenAI supports deeper cognitive engagement, promoting critical analysis and a better comprehension of complex topics. We also identified terms like "idea generation," "problem-solving," and "comparing," which reflect an appreciation for GenAI in facilitating creative thinking and intellectual comparison. This aids in developing innovative research approaches and solutions. Overall, these terms indicate that GenAI is seen not just as a convenient tool but also as a catalyst for enhancing intellectual rigor and innovation in the research process.

The analysis of negative terms (see Fig. 1.5) highlighted key concerns faced by researchers, such as "ethics" and "accuracy." This indicates a strong focus on the moral and factual implications of using AI in academic work. The terms "responsibility" and "declaration" suggest a need for transparency and accountability when using AI tools. The presence of "bias" points to worries about the fairness and neutrality of AI outputs, while "creativity" reflects concerns that over-reliance on AI may stifle original thought. Other terms like "repetition," "cost," and "reliance" indicate anxieties about AI's limitations in generating novel insights, the potential financial burden, and the risk of researchers becoming overly dependent on these tools. Lastly, "environment" suggests concerns regarding the sustainability and broader ecological impact of using AI technologies. Together, these keywords underscore the multifaceted apprehensions surrounding the integration of GenAI

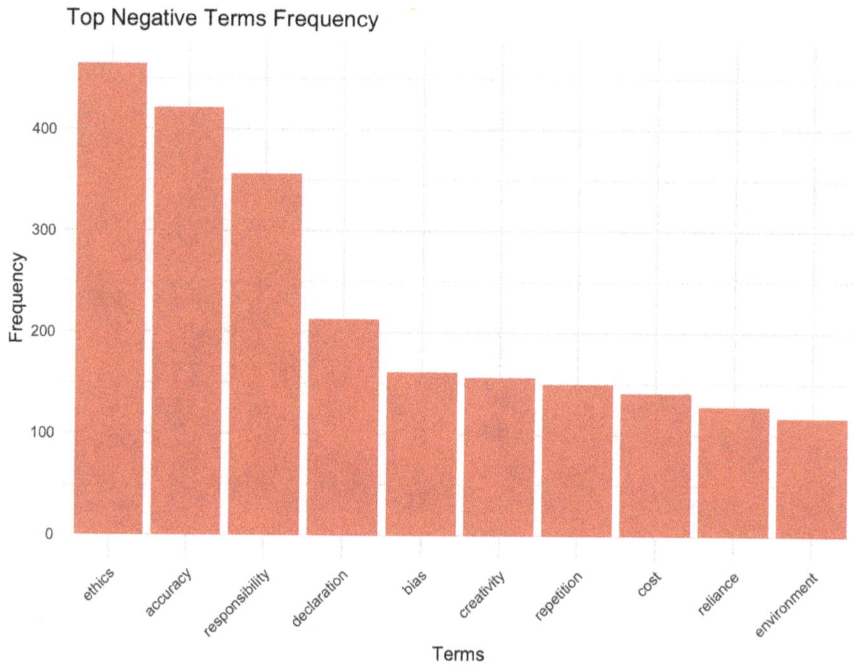

Fig. 1.5 Top negative terms on the use of GenAI in research

into academic research, particularly regarding its ethical, practical, and long-term implications.

4 Collective Guidance on GenAI Use in Research

The use of GenAI across disciplines can bring different challenges when it comes to formulate future policies and usage strategies. This is largely reflected in each discipline, and the research needs of students. Meanwhile, AI is viewed by individuals doing research as an enabler of new methods, processes, management, and evaluation (Alasadi & Baiz, 2023; Dwivedi et al., 2023). Based on our review of large volume of previous studies, our seeding initiatives, across discipline workshops, and close consultation with educational partners across the UK, we found that the majority of students enrolled in research programmes across all disciplines are using GenAI tools for idea generation, planning of the literature, identify research methods and procedures, perform data analysis and interpretation.

Table 1.1 compares the various tools, skills, and practices students might develop through the use of GenAI in their research. During the idea generation phase, tools like ChatGPT and Jasper have been noted for their ability to help students brainstorm by suggesting potential research questions and offering related perspectives.

Table 1.1 Key tools, practices and skills of using GenAI in research

Stage	Use	Tools	Practices	Skills
Idea generation	Breakdown of logics and ideas, generate relevant ideas, repropose closely related ideas, provide alternatives, compare between solutions and link them to the searched problem	ChatGPT, Jasper, IdeaFlow	Brainstorming research questions and identifying knowledge gaps	Critical thinking, information seeking, and adaptability
Planning of the literature	Identify search strategy, build keywords, article filtering criteria, suggest connections between studies, compare between studies, identify gap in knowledge, suggest research questions	Elicit, Iris.ai, Connected Papers	Quickly review literature and summarize findings	Analytical reading, synthesis of summaries
Identify research methods and procedures	Support in sample identification, recruitment methods, protocol, and procedure	SciNote, GPT models for drafting	Formulate hypotheses, suggest research frameworks, sample sizes, and methodologies	Research methodology, critical analysis
Data analysis and interpretation	Selection of data analysis methods, visualisation, writing style and interpretation of statistical data	OpenAI Codex, Python, R, DALL-E, ChatGPT, WriteSonic, Grammarly	Plan data collection and analysis methods, simulate potential outcomes	Data literacy, statistical reasoning

Regular use of these tools is thought to aid students in developing critical thinking skills, essential for identifying research gaps or forming hypotheses. Tools such as Elicit and Iris.ai, used in the planning of literature reviews, can assist students in summarising large amounts of academic work, potentially simplifying the process of identifying relevant research and emerging trends. The development of analytical skills is seen in students' ability to synthesise information efficiently, though the full impact of using GenAI for literature reviews is still not fully understood in the current literature.

Furthermore, GenAI tools like SciNote or GPT can be effective in supporting students identify relevant research methods and procedures that are related to their subject area and sample characteristics. These tools support students by assessing them in drafting research frameworks, suggesting methodologies, and facilitating an understanding of key assumptions related to the subject matter. In terms of data analysis and interpretation, GenAI tools such as OpenAI Codex can recommend suitable data analysis techniques and visualisations for specific research types. The potential advantage of these tools at this stage lies in improving statistical reasoning, as students are expected to select and apply appropriate AI-generated recommendations while avoiding over-reliance on AI outputs. Nevertheless, this process demands that students make their own decisions and critically link these to existing recommendations from their literature review.

5 How AI Is Supporting Research Skills Development?

AI is transforming how researchers approach their work, from literature review, and hypothesis generation to data analysis. Khalifa and Albadawy (2024) carried out a systematic review to examine how AI is utilised and its impact on academic writing and research. Their research revealed that AI has improved academic writing and research across six key areas: idea generation, content structuring, literature synthesis, data management, editing, and ethical compliance, as illustrated in Fig. 1.6. Both UNESCO (2023) and the European Commission (2024) have acknowledged the growing trend of using General AI to aid academic research. They have provided detailed guidelines to support academics in effectively utilising GenAI for research purposes. In the following, we delved into the specific ways AI is fostering research skills development.

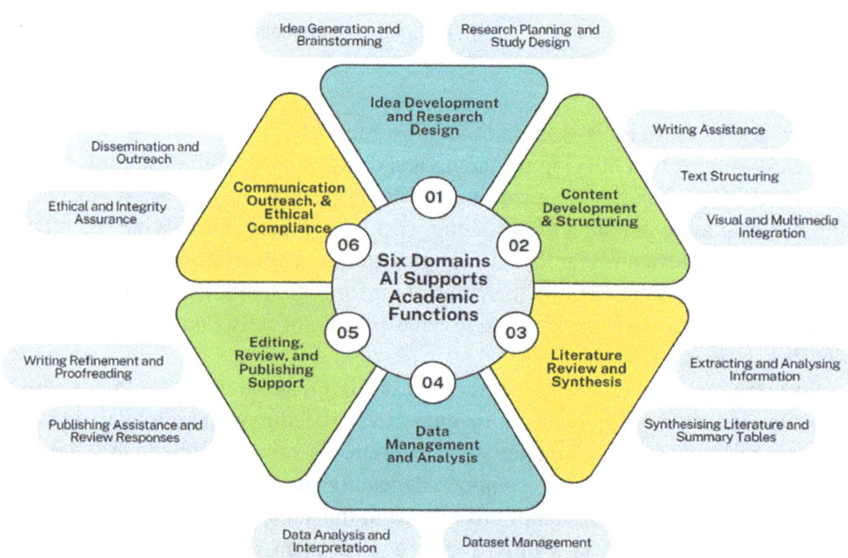

Fig. 1.6 The six domains where AI can improve academic functions (Khalifa & Albadawy, 2024)

5.1 *Facilitation of Literature Review and Information Synthesis*

One of the most significant contributions of AI to research skills development is in the area of literature review and information synthesis. The exponential growth of scientific publications makes it increasingly challenging for researchers to stay abreast of developments in their field. AI-powered tools are helping to address this challenge, while also fostering important research skills (Wagner et al., 2022).

Natural Language Processing (NLP) algorithms can rapidly analyse vast numbers of scientific papers, extracting key information, identifying trends, and even suggesting connections between seemingly disparate studies. These tools are not just saving time; they're changing how researchers approach the literature review process. AI tools that blend large language models (LLMs) with academic databases, such as Elicit and Semantic Scholar, are extensively utilised to assist researchers in finding relevant literature. Some conversational AI tools, such as the GPT-4 model, allow for the use of retrieval-augmented LLMs that access external documents, thereby enhancing accuracy and reducing factual errors (Bubeck et al., 2023). AI-driven literature review tools such as Connected Papers, Inciteful and Research Rabbit can generate knowledge graphs, visually representing the relationships between different concepts, authors, and publications in a field (Aaron, 2021). By interacting with these knowledge graphs, researchers can develop skills in conceptual mapping and interdisciplinary thinking. Students can also learn to identify connections across different areas of study and identify gaps in existing research more effectively. Furthermore, AI tools, for example, Scispace and ChatGPT, that can summarise complex scientific papers, and even provide detailed explanations of the concepts for students to quickly grasp the essence of numerous studies (Semrl et al., 2023). This capability is precious for interdisciplinary research, where researchers often need to quickly develop to speed on areas outside their primary expertise. By working with these tools, researchers are honing their skills in rapid information assimilation and synthesis.

However, it's crucial to note that these tools are not replacing the need for careful, critical reading. Instead, they're allowing researchers to approach literature review more strategically, focusing their in-depth reading on the most relevant and promising papers (Khlaif et al., 2023). This strategic approach is fostering the development of more sophisticated information literacy skills, as researchers learn to navigate and make sense of the vast sea of available information more effectively.

5.2 *Enhancement of Data Analysis and Pattern Recognition Capabilities*

Another area where AI can significantly support research skills development is in data analysis and pattern recognition. As datasets grow increasingly large and complex, traditional analytical methods often fall short. Machine learning algorithms excel at processing vast amounts of data and identifying patterns that might be imperceptible to human researchers (Sharma et al., 2024). Research has also indicated that AI-driven software and tools, equipped with automated guiding strategies, can produce more precise data analysis by quickly identifying key traits compared to conventional data analysis methods (Zhai et al., 2023).

Advanced AI systems can help researchers analyse data across multiple dimensions in order to uncover hidden patterns and relationships within datasets. This capability is particularly valuable in fields like genomics, climate science, and social network analysis, where the interplay of numerous variables can be challenging to discern through conventional methods. For instance, in bioinformatics, AI algorithms can analyse gene expression data to identify potential biomarkers for cardiovascular diseases, a task that would be prohibitively time-consuming and complex for human researchers alone (DeGroat et al., 2024). Similarly, in climate science, AI models can process data from multiple sources—satellite imagery, weather stations, and historical records—to detect subtle patterns in climate change that might otherwise go unnoticed (Satish et al., 2023).

GenAI has proven to be a valuable tool in supporting researchers in both qualitative and quantitative data analysis. Perkins and Roe (2024b) demonstrated how ChatGPT could be used for thematic analysis that can contribute to improved coding efficiency and aiding in initial data exploration. Similarly, Perkins and Roe (2024b) found that ChatGPT significantly aids inductive thematic analysis by enabling a process that includes separate human-driven and GenAI-assisted coding. While ChatGPT's utility in data analysis was acknowledged, issues such as hallucination and irreproducibility in outputs were reported. This as a result emphasises the need for researchers to verify and critically assess the results. Furthermore, GenAI has shown greater accuracy in supporting quantitative data analysis. Tools integrated with popular statistical software have enabled more sophisticated analyses, streamlining processes from data cleaning and identifying outliers or missing data patterns to suggesting appropriate statistical models based on data characteristics (Perkins & Roe, 2024a). This integration not only simplifies the analytical process for researchers but also alleviates concerns about the reproducibility of results.

By working with these AI tools, researchers are likely to develop enhanced skills in data interpretation and pattern recognition. They're learning to ask more nuanced questions of their data, formulate more complex hypotheses, and think more critically about the patterns and relationships uncovered by AI analysis. Moreover, as researchers become more adept at working with AI-driven data analysis tools, they're developing a deeper understanding of the strengths and limitations of these technologies. This growing AI literacy is becoming an essential skill in itself, enabling researchers to more effectively leverage these powerful tools while maintaining a critical perspective on their outputs.

5.3 Assistance in Research Design and Hypothesis Generation

AI is also playing an increasingly important role in research design and hypothesis generation, two critical aspects of the research process. By analysing patterns in existing research and combining elements from different studies, some advanced AI systems can suggest research questions and novel hypotheses for researchers to explore (Alshater, 2022). For instance, in drug discovery research, AI algorithms can analyse data on molecular structures, biological pathways, and existing drugs to propose new compounds that might be effective against a particular disease (Chen et al., 2023).

While the researcher still needs to evaluate these suggestions critically, this AI-assisted approach is fostering more creative and wide-ranging hypothesis generation. In terms of experimental design, AI tools can help researchers optimise their methodologies. Machine learning algorithms can simulate numerous experimental scenarios, helping researchers identify the most promising approaches before they even step into the lab (Ma et al., 2024). This capability is particularly valuable in fields where experiments are costly or time-consuming, such as particle physics or clinical trials. Moreover, the latest GenAI tools can generate preliminary research ideas, methodologies, and expected outcomes with simple prompts like "Based on the research gap, generate 10 topical research questions for topic X" (UNESCO, 2023). These tools provide foundational concepts that can inform the overall research plan.

By engaging with these AI tools, researchers are developing more sophisticated skills in hypothesis formulation and experimental planning. They're learning to think more systematically about the variables in their experiments and to consider a wider range of possibilities in their hypotheses. Moreover, the process of critically evaluating AI-generated hypotheses and experimental designs is itself fostering important research skills. Researchers are developing a more nuanced understanding of what makes a good hypothesis or a well-designed experiment, as they learn to discern the strengths and weaknesses of AI-generated suggestions.

5.4 Role of AI in Academic Writing

While AI in academic writing is often associated with automating tasks like grammar checking and citation formatting, its true impact lies in fostering higher-order writing skills such as critical analysis, argumentation, and stylistic refinement (Khalifa & Albadawy, 2024). This might seem counterintuitive, but it stems from how AI is transforming the nature of academic writing. As AI systems take over more routine aspects of writing—like basic editing, reference management, and initial literature searches—writers are freed to focus on more complex, creative aspects of their work. They're spending more time developing nuanced arguments, synthesising diverse sources, and critically examining the implications of their research.

Furthermore, the use of AI writing tools like Grammarly, which specialises in enhancing writing skills, is fostering advanced writing abilities. These tools offer suggestions for impeccable English grammar structures and vocabulary variations, making them highly beneficial for non-English-speaking academics aiming to correct grammatical errors. Utilising AI writing tools is not only enhancing advanced writing skills but also effectively improving the writing process and boosting productivity (Golan et al., 2023). However, this integration poses challenges, particularly for junior academics with limited research experience and subject knowledge. Researchers must carefully consider how to present their ideas to AI systems, interpret AI-generated suggestions judiciously, and seamlessly blend AI-assisted content with their original thoughts and analyses. Argumentation skills or critical analysis skills are also being enhanced through AI integration. Writers are learning to construct more robust arguments by leveraging AI's ability to quickly identify counterarguments and potential weaknesses in logic (Hadan et al., 2024). For instance, AI-powered research tools such as Elicit and Consensus, can surface conflicting viewpoints from a vast array of sources, challenging writers to address these perspectives and strengthen their own positions.

A growing number of publishers are recognising that AI can be utilised to enhance the readability and grammatical correctness of articles. For example, the Academy of Management (n.d.) has specified in its AI policy that while authors are permitted to use AI for spelling and grammar checks, they must inform the publisher about the sections of the manuscript where AI was used. Similarly, Elsevier's (n.d.) AI policy indicates that AI can be employed to refine the language of a document, but it should not be used to create the scientific and pedagogical content of the paper. In essence, AI in academic writing is not replacing the need for human skills but rather elevating the role of the writer, pushing authors to engage more deeply with their subject matter and produce more refined scholarly works. Authors are responsible for ensuring that the final output of their writing is clear and accurate, effectively communicating the research in a concise and precise manner. Our view is that the most successful researchers of the future will likely be those who can effectively combine their domain expertise with AI literacy to push the boundaries of human knowledge while maintaining a critical, ethical perspective on their use.

6 Responsible AI Integration

The integration of AI into research processes offers unprecedented opportunities for advancing knowledge and accelerating scientific discovery. However, this integration also brings some ethical considerations and practical challenges that researchers must understand. The followings are the main considerations.

6.1 Ethical Considerations

- Data Privacy:
- One of the foremost ethical concerns in AI-driven research is data privacy. AI systems often require vast amounts of data to function effectively, and in many research contexts, this data may include sensitive personal information. Researchers must grapple with questions of data ownership, consent, and the potential for re-identification of anonymised data. The challenge lies in balancing the need for comprehensive datasets with the ethical imperative to protect individual privacy. Techniques such as differential privacy and federated learning are emerging as potential solutions, allowing researchers to gain insights from data without directly accessing individual records (UNESCO, 2023). However, implementing these techniques can be complex and may impact the accuracy of results, presenting researchers with difficult trade-offs. Moreover, as AI systems become more sophisticated, there's an increasing risk of unintended data disclosures. For instance, an AI model trained on anonymised data might inadvertently learn patterns that could lead to re-identification of individuals. Researchers must be vigilant about these risks and implement robust safeguards to protect privacy throughout the research process (Perkins & Roe, 2024a).
- Bias in AI Algorithms:
- Another critical ethical consideration is the potential for bias in AI algorithms. AI systems learn from the data they're trained on, and if this data contains historical biases or is not representative of diverse populations, the resulting algorithms can perpetuate or even amplify these biases (Hacker et al., 2024). In research contexts, biased AI systems can lead to skewed results, misrepresentation of certain groups, and ultimately, flawed conclusions (Afrin et al., 2022). This is particularly concerning in fields like healthcare, social sciences, and policy research, where biased outcomes could have far-reaching societal implications. Addressing algorithmic bias requires a multifaceted approach. Researchers must critically examine their data sources, actively seek diverse and representative datasets, and implement bias detection and mitigation techniques in their AI models. However, identifying and correcting for all potential biases is a complex task that requires ongoing vigilance and interdisciplinary collaboration.

- Transparency and Explainability:
- The "black box" nature of many advanced AI systems, particularly deep learning models, poses significant challenges for transparency in research. When AI is used to analyse data or generate insights, it's crucial that researchers understand and explain how these results were obtained. The lack of explainability in AI systems or tools can undermine the scientific principle of reproducibility and make it difficult to validate research findings. Furthermore, in fields where AI-driven decisions have significant consequences, such as medical diagnosis or criminal justice research, the inability to explain how an AI reached its conclusion raises serious ethical concerns.

6.2 *Responsible AI Integration in Research*

AI techniques and tools are evolving at an unprecedented rate, making it difficult for researchers, especially those without a background in computer science, to stay informed about the latest developments and best practices. This rapid evolution also means that ethical guidelines and regulatory frameworks often lag behind technological capabilities (UNESCO, 2023). Researchers may find themselves in unexplored ethical territory, having to make decisions about the use of AI technologies before clear guidelines have been established. To address this challenge, there's a growing need for ongoing education and interdisciplinary collaboration. Researchers need access to resources and training that can help them understand the ethical implications of new AI technologies and how to implement them responsibly (Perkins & Roe, 2024b).

- Balancing AI Capabilities with Human Expertise:
- Another significant challenge lies in striking the right balance between leveraging AI capabilities and maintaining the critical role of human expertise in research. While AI can process vast amounts of data and identify patterns beyond human capability, it lacks the nuanced understanding, creativity, and ethical judgment that human researchers bring to the table. Solid research continues to depend on researchers' deep subject knowledge, thorough analysis, and interpretation. AI cannot substitute human intelligence when it comes to critical thinking and the creation of new knowledge. There's a risk that over-reliance on AI could lead to a devaluation of human expertise or a reluctance to question AI-generated results. This may particularly affect junior researchers by depriving them of valuable learning experiences through trial and error. Researchers must cultivate a mindset that views AI as a tool to augment human capabilities rather than replace them. This requires developing new skills in interpreting and critically evaluating AI outputs, as well as designing research methodologies that effectively combine AI and human insights (European Commission, 2024).

- Ensuring Equitable Access and Preventing AI Divide:
- As AI becomes increasingly integral to cutting-edge research, there is a risk of exacerbating existing inequalities in the research landscape. Advanced AI systems often require significant computational resources and technical expertise, which may not be equally available to all researchers or institutions. Additionally, researchers who have access to more advanced versions of GenAI, such as ChatGPT 4, often experience enhanced productivity and effectiveness. This potential "AI divide" could lead to a concentration of AI-driven research capabilities in well-funded institutions or technologically advanced countries, potentially skewing the direction of global research and limiting diverse perspectives. Addressing this challenge requires concerted efforts to democratise access to AI technologies in research. This might involve developing more accessible AI tools, promoting open-source AI initiatives, and investing in AI education and infrastructure in underrepresented communities and developing countries.

In conclusion, responsible AI integration in research presents a complex landscape of ethical considerations and practical challenges. Addressing issues of data privacy, algorithmic bias, and transparency, while also navigating the rapid pace of technological change and ensuring equitable access, requires a concerted effort from the entire research community. The key to responsible AI integration lies in maintaining a balance—leveraging the power of AI to enhance research capabilities while steadfastly upholding ethical principles and the irreplaceable value of human insight and oversight. As we continue to explore the frontiers of AI-driven research, ongoing dialogue, interdisciplinary collaboration, and a commitment to ethical practice will be crucial in realising the full potential of AI while safeguarding the integrity and societal impact of research.

7 Summary

Efforts to support researchers' independence in using AI are still in their early stages in higher education. This is mainly due to the constantly changing nature of AI technology, which has led to shifts in educational policies and practices. As a result, there's considerable uncertainty about how best to integrate AI tools into research. To make the most of these developments, universities need to help researchers adapt to and use GenAI in a way that's both ethical and effective. There's also a growing interest among universities to define the role of AI in research development and policy. However, there's a noticeable gap between what students expect and what institutions prioritise, particularly when it comes to research outcomes and employability. This disconnect has made things challenging for both students and their supervisors, especially as timely supervision and mentoring are vital. The mismatch between expectations has created confusion around AI research practices, which can lead to misunderstandings and knowledge gaps. This situation can negatively impact students' research results and their ability to meet institutional objectives. While GenAI could help students, especially those who lack knowledge in certain areas, integrating it comes with challenges. These include a lack of resources, guidance, and clear procedures for using GenAI responsibly in research.

References

Aaron, T. (2021). 3 new tools to try for literature mapping. *Meium*. https://aarontay.medium.com/3-new-tools-to-try-for-literature-mapping-connected-papers-inciteful-and-litmaps-a399f27622a

Academy of Management. (n.d.). *AOM Artificial Intelligence Policy*. Retrieved September 3, 2024, from https://aom.org/research/publishing-with-aom/aom-artificial-intelligence-policy

Afrin, F., Hamilton, M., Thevathyan, C., & Majrashi, K. (2022). Investigating perceptions of AI-based decision making in student success prediction. *Lecture notes in computer science (including subseries lecture notes in artificial intelligence and lecture notes in bioinformatics), 13356 LNCS* (pp. 315–319). https://doi.org/10.1007/978-3-031-11647-6_60

Alasadi, E. A., & Baiz, C. R. (2023). Generative AI in education and research: Opportunities, concerns, and solutions. *Journal of Chemical Education, 100*(8), 2965–2971. https://doi.org/10.1021/ACS.JCHEMED.3C00323/ASSET/IMAGES/MEDIUM/ED3C00323_0002.GIF

Al-Samarraie, H., Sarsam, S. M., Ibrahim Alzahrani, A., Chatterjee, A., & Swinnerton, B. J. (2024). Gender perceptions of generative AI in higher education. *Journal of Applied Research in Higher Education* (ahead-of-print). https://doi.org/10.1108/JARHE-02-2024-0109/FULL/XML

Alshater, M. M. (2022). Exploring the role of artificial intelligence in enhancing academic performance: A case study of ChatGPT. *SSRN Electronic Journal*. https://doi.org/10.2139/SSRN.4312358

Andersen, J. P., Degn, L., Fishberg, R., Graversen, E. K., Horbach, S. P. J. M., Schmidt, E. K., Schneider, J. W., & Sørensen, M. P. (2024). Generative artificial intelligence (GenAI) in the research process – A survey of researchers' practices and perceptions. *SocArxiv*, 1–34. https://doi.org/10.31235/OSF.IO/83WHE

Bubeck, S., Chandrasekaran, V., Eldan, R., Gehrke, J., Horvitz, E., Kamar, E., Lee, P., Lee, Y. T., Li, Y., Lundberg, S., Nori, H., Palangi, H., Ribeiro, M. T., & Zhang, Y. (2023). arXiv-2303.12712v5.

Chen, W., Liu, X., Zhang, S., & Chen, S. (2023). Artificial intelligence for drug discovery: Resources, methods, and applications. *Molecular Therapy - Nucleic Acids, 31*, 691–702. https://doi.org/10.1016/J.OMTN.2023.02.019

DeGroat, W., Abdelhalim, H., Patel, K., Mendhe, D., Zeeshan, S., & Ahmed, Z. (2024). Discovering biomarkers associated and predicting cardiovascular disease with high accuracy using a novel nexus of machine learning techniques for precision medicine. *Scientific Reports, 14*(1), 1–13. https://doi.org/10.1038/s41598-023-50600-8

Dwivedi, Y. K., Kshetri, N., Hughes, L., Slade, E. L., Jeyaraj, A., Kar, A. K., Baabdullah, A. M., Koohang, A., Raghavan, V., Ahuja, M., Albanna, H., Albashrawi, M. A., Al-Busaidi, A. S., Balakrishnan, J., Barlette, Y., Basu, S., Bose, I., Brooks, L., Buhalis, D., Wright, R., et al. (2023). Opinion paper: "So what if ChatGPT wrote it?" Multidisciplinary perspectives on opportunities, challenges and implications of generative conversational AI for research, practice and policy. *International Journal of Information Management, 71*, 102642. https://doi.org/10.1016/J.IJINFOMGT.2023.102642

Elsevier. (n.d.). *The use of AI and AI-assisted technologies in writing for Elsevier*. Retrieved September 3, 2024, from https://www.elsevier.com/en-gb/about/policies-and-standards/the-use-of-generative-ai-and-ai-assisted-technologies-in-writing-for-elsevier

European Commission. (2024). *Living guidelines on the responsible use of generative AI in research*. http://data.europa.eu/eli/dec/2011/833/oj

Golan, R., Reddy, R., Muthigi, A., & Ramasamy, R. (2023). Artificial intelligence in academic writing: A paradigm-shifting technological advance. *Nature Reviews Urology, 20*(6), 327–328. https://doi.org/10.1038/s41585-023-00746-x

Hacker, P., Zuiderveen Borgesius, F., Mittelstadt, B., Wachter, S., Engel, A., & Hammer, S. (2024). *Generative discrimination: What happens when generative AI exhibits bias, and what can be done about it.* https://arxiv.org/abs/2407.10329v1.

Hadan, H., Wang, D. M., Hadi, R., Tu, J., Zhang-Kennedy, L., & Nacke, L. E. (2024). The great AI witch hunt: Reviewers' perception and (mis)conception of generative AI in research writing. *Computers in Human Behavior: Artificial Humans, 1.*

Khalifa, M., & Albadawy, M. (2024). Using artificial intelligence in academic writing and research: An essential productivity tool. *Computer Methods and Programs in Biomedicine Update, 5*, 100145. https://doi.org/10.1016/j.cmpbup.2024.100145

Khlaif, Z. N., Mousa, A., Hattab, M. K., Itmazi, J., Hassan, A. A., Sanmugam, M., & Ayyoub, A. (2023). The potential and concerns of using AI in scientific research: ChatGPT performance evaluation. *JMIR Med Educ, 9*, E47049. https://Mededu.Jmir.Org/2023/1/E47049. https://doi.org/10.2196/47049

Ma, J., Cao, B., Dong, S., Tian, Y., Wang, M., Xiong, J., & Sun, S. (2024). MLMD: A programming-free AI platform to predict and design materials. *Npj Computational Materials, 10*(1), 1–12. https://doi.org/10.1038/s41524-024-01243-4

Perkins, M., & Roe, J. (2024a). *Generative AI tools in academic research: Applications and implications for qualitative and quantitative research methodologies.*

Perkins, M., & Roe, J. (2024b). The use of generative AI in qualitative analysis: Inductive thematic analysis with ChatGPT. *Journal of Applied Learning and Teaching, 7*(1), 390–395. https://doi.org/10.37074/JALT.2024.7.1.22

Satish, M., Prakash, B. S. M., Kumar, P. P., Devi, S., & Reddy, K. P. (2023). Artificial intelligence (AI) and the prediction of climate change impacts. *5th IEEE international conference on cybernetics, cognition and machine learning applications, ICCCMLA 2023* (pp. 660–664). https://doi.org/10.1109/ICCCMLA58983.2023.10346636

Semrl, N., Feigl, S., Taumberger, N., Bracic, T., Fluhr, H., Blockeel, C., & Kollmann, M. (2023). AI language models in human reproduction research: Exploring ChatGPT's potential to assist academic writing. *Human Reproduction, 38*(12), 2281–2288. https://doi.org/10.1093/HUMREP/DEAD207

Sharma, D., Chaudhary, G., Rajput, K., Singhal, A., & Sharma, T. K. (2024). Utilizing artificial intelligence for advance data science and analysis. *2024 2nd international conference on disruptive technologies, ICDT 2024* (pp. 992–998). https://doi.org/10.1109/ICDT61202.2024.10489078

UNESCO. (2023). Guidance for generative AI in education and research. *Guidance for Generative AI in Education and Research.*

Wagner, G., Lukyanenko, R., & Paré, G. (2022). Artificial intelligence and the conduct of literature reviews. *Journal of Information Technology, 37*(2), 209–226. https://doi.org/10.1177/02683962211048201

Zhai, K., Yousef, M. S., Mohammed, S., Al-Dewik, N. I., & Qoronfleh, M. W. (2023). Optimizing clinical workflow using precision medicine and advanced data analytics. *Processes 2023, 11*(3), 939. https://doi.org/10.3390/PR11030939.

Chapter 2
Student Perspectives on the Impact of GenAI

Cassandra Lewis, Paula Funnell, Nick Fisher, and Pedro Elston

Abstract This chapter explores student perspectives on the impact of Generative AI (GenAI) in higher education. By examining narratives across various disciplines—healthcare, business, human resources, law, English literature, and computing—it seeks to understand the benefits and challenges GenAI presents to students, as well as its broader implications for higher education and future careers. Qualitative data from student surveys from various fields was gathered to showcase student perspectives through a narrative analysis. This approach captures diverse views on how GenAI might affect learning and research, which can offer insights into students' experiences in an AI-influenced academic environment. The findings highlight that students are keenly aware of GenAI's impact on their education, often more so than their educators or instructors. While some students are optimistic about GenAI's potential, others express concern over the ethics of its use, its effects on traditional learning, and implications on their professions. This chapter highlights the necessity of including student voice in discussions about GenAI's role in higher education. As primary stakeholders, their insights are vital in shaping educational policies and practices in response to GenAI's growing presence. The findings also contribute to ongoing debates around the role of higher education in preparing students for the changing requirements of future careers in an increasingly AI-driven world.

Keywords GenAI · Student perspectives · Employability · Technology · Skills · Literacies

---

C. Lewis (✉) · P. Funnell · N. Fisher · P. Elston
Queen Mary University of London, London, UK
e-mail: cassandra.lewis@qmul.ac.uk

X. Zhou, H. Al-Samarraie (eds.), *Institutional guide to using AI for research*, Advances in Artificial Intelligence in Education 2,
https://doi.org/10.1007/978-3-031-94809-1_2

1 Introduction

The global higher education arena has been afforded incredible benefits but also challenges with the arrival of digital learning systems and intelligent technologies, but nothing as significant or as transformative as the rise of generative artificial intelligence (GenAI) (Crompton & Burke, 2023; Yusuf et al., 2024). It is also without question that in 2024, the job market was calling for AI talent. While GenAI may improve productivity, there is the potential for significant levels of job loss and transformation through automation of traditional processes across multiple sectors, particularly creative industries (Malik et al., 2023; Zirar et al., 2023). This includes Art and Design, Illustration, Photography, but also areas of Finance, Entertainment and Legal Services (Dwivedi et al., 2023).

Students must be at the forefront of discussions around GenAI use, and their viewpoints are pivotal to navigating changes in policy and co-creation of the future of higher education (Chan & Hu, 2023; Johnston et al., 2024). Indeed, students are well-versed, even better-versed (Coffey, 2023) than staff in appreciating the impact this has on their education and their future careers, particularly how this may differ between disciplines and industries (Malik et al., 2023). This chapter aims to explore perspectives and experiences through student narratives across the disciplines.

1.1 GenAI in Higher Education

There has been an 'AI boom' in the last 5 years, driven by the rapid expansion of Large Language Models (LLMs) such as OpenAI's ChatGPT in 2022, and subsequently, Google Gemini and Microsoft Copilot, among many others (Crompton & Burke, 2023). Newer versions of AI models are being integrated with email software, word processors and spreadsheets, and even maps. It is our students who will be most impacted by the transformative process of GenAI applications, within their studies and upon graduation into a new AI-integrated age. The unparalleled development and proliferation of these models have brought social, ethical, economic and legal issues to the fore (Malik et al., 2023; Yusuf et al., 2024), stemming from concerns about data security and privacy, intellectual theft, copyright and plagiarism, and a perception of accuracy, transparency, accessibility and inequity (D'Agostino, 2023).

Despite the rapid global adoption of ChatGPT (reaching one million users within the first 5 days), there are divergent viewpoints and perspectives on the effect GenAI will have on higher education. Bearman et al. (2023) highlight the contrasting discourses of AI as either a 'utopian' asset which could revolutionise higher education, or a 'dystopian' threat to knowledge acquisition and learning. Indeed, Oxford and Cambridge universities initially moved to ban ChatGPT in 2023 (UNESCO, 2023b). Since then, a pro-innovation approach has been advocated by key authorities in the United Kingdom, including the Department for Education (2023), Joint Information

Systems Committee (JISC) (2023), Quality Assurance Agency (2023a, 2023b), and The Russell Group of Universities (2023). All these organisations call upon higher education institutions to work with AI systems and integrate this emerging technology into curricula. Such integration requires sophisticated policies which reflect cultural awareness and expectations (Yusuf et al., 2024), promote academic honesty, ethics and integrity, whilst leveraging the benefits afforded to students and educators (Yusuf et al., 2024).

Students and educators must remain cognizant of the human relationship with these new technologies, which may potentially impact on student motivation, levels of stress, energy and potentially even their identities (Kuadey et al., 2023; Thorpe, 2023). Indeed, Kramm and McKenna (2023) address the question of what is "higher education *really* for?" and what might be the impact on the roles of the student and educator? There is a need to develop national (or even global) policy on AI-use in higher education, and for ubiquitous training in AI literacy for students and staff; to re-educate on topics such as academic integrity and misconduct, alongside a rapid review of the validity of current assessments and assessment processes (Malik et al., 2023; Quality Assurance Agency, 2023a).

1.2 Job Reduction or Evolution?

The impact of AI on the job market is predicted to be significant, with potentially 85 million displaced jobs by 2025. Reassuringly, however, the International Monetary Fund predicted the emergence of 97 million future jobs (Zahidi, 2020), specifically emerging roles in AI and creation careers such as social media, media management and content writing. Research in 2024 by Microsoft and LinkedIn (Kimbrough, 2024) suggested the demand for AI talent rose by 323% on LinkedIn, with a 142-fold increase in AI literacy skills added to LinkedIn profiles across diverse industries, particularly in the creative arts e.g. graphic design, content writing and marketing. According to this data, the fastest growing industries in the US for AI tech talent included Consumer Services (35%), Education (29%), Entertainment Providers (26%), Administrative and Support Services (25%) and Accommodation and Food Services (25%). There was a clear realisation that AI skills were essential to power the future success of organisations and indeed entire industries. Amazon Web Services (2023), in collaboration with Access Partnership, surveyed over 4000 US employees and institutions. Their research concluded that while 92% of employers thought AI would benefit their IT departments, 85% felt the benefits were expected to extend to Sales and Marketing, and 78% to Human Resources. Furthermore, it was predicted that salaries could increase by as much as 35% for those with AI expertise, particularly in IT and Sales and Marketing positions. There are benefits afforded to both employers and employees, including improved workload management, enhanced creativity, and opportunities to focus on more important work, with predicted productivity increases of 47%. In addition, there seems a clear demand for training in AI literacy and proficiency to remain relevant and

Table 2.1 Top ten core skills required by 2030 (World Economic Forum, 2023)

Rank	Core skills
1	Creative thinking
2	Analytical thinking
3	Technological literacy
4	Curiosity and lifelong learning
5	Resilience, flexibility, and agility
6	Systems thinking
7	AI and big data
8	Motivation and self-awareness
9	Talent management
10	Service orientation and customer service

competitive in the job market (Attewell, 2023). According to the World Economic Forum (2023), cognitive skills evolution, self-efficacy and problem-solving are some of the key skillsets required by our employees of the future. The following top ten core skills were predicted to be of most importance by 2030 (See Table 2.1):

While these core skills remain relatively uniform across industries, specific skill-sets vary in importance across the different sectors. Certain careers, particularly in Care, Personal Services and Wellbeing are seeing the most dramatic increase in demand for technological skills. Despite this growing trend, only 25% of employers were shown to offer AI training programs (World Economic Forum, 2023) and few were investing in AI tools for their employees. Indeed 82% of employers were unaware of how to implement such training programmes and many (75%) reported difficulty in recruiting qualified individuals, compounding this skill gap (Amazon Web Services, 2023).

In 2024, therefore, new graduates were faced with upskilling independently, with 75% of workers adopting self-bought AI tools to leverage the latest AI tech (Kimbrough, 2024). Amazon Web Services (2023) suggested that by 2028, 88% of workers expect to use AI in their daily jobs, with 25% expecting to use AI "extensively" (for over 60% of their tasks). Clearly training and support is needed, as business and process transformation will require adoption, innovation and continuous upskilling of the workforce to remain both agile and competitive. Higher education institutions also need to rapidly evolve curricula and provide AI literacy training to reflect workplace transformation and adequately prepare students for AI-enhanced employment.

Therefore, the aim of this qualitative study was to explore the experiences, perspectives and perceptions of postgraduate students across the disciplines, on the use of GenAI in higher education and research, and the impact on their future careers.

2 Methods

This study employed a qualitative methodology to explore the lived experiences of postgraduate students from various Schools at Queen Mary University London (QMUL) in 2024. Students were selected through purposive sampling owing to their previous involvement in QMUL-funded projects relating to the use of GenAI and digital technologies in higher education. As students with prior interest and experience, this inclusion criteria served to select not only key stakeholders but also knowledge experts in the field. By exploring the disruption through these students' lens, we hope to showcase a range of challenges and opportunities facing students in an emerging AI-driven world. Seven students were approached, and all volunteered to complete an online survey exploring their views on the use and impact of AI in education, research, and future careers. Their perspectives came from various academic backgrounds, including students from MA in English Literature (A1), MSc in International Business (A2), MSc in Commercial and Corporate Law (A3), BDS Dentistry (A4), PhD in Healthcare (A5), MSc in International Human Resources Management (A6), and MSc in Computing and Information Systems (A7). Notations' A1–A7 are used to refer to the recruited students when interpreting their perspectives on the use and impact of GenAI. The authors conducted a narrative analysis, which involves telling and re-telling life stories to co-construct knowledge and describe the student viewpoint. The authors have previously collaborated on QMUL-funded projects relating to GenAI literacy and capability, with a focus on co-creation of an e-resource to support students in the ethical use of GenAI in their studies. Limitations of these findings are acknowledged, and stem from potential confirmation bias and courtesy bias, as some students were involved in previous academic collaborations and were known to the authors.

3 Results

This section will outline the findings of the narrative analysis, exploring student perspectives on the implications of GenAI across their respective disciplines. Three key themes were identified, relating to the risks and benefits afforded by GenAI to: (1) higher education, (2) research and (3) future careers. These are presented below, illustrated by verbatim quotes from each of the seven students (A1–A7).

3.1 *Theme 1: Impact of GenAI on Higher Education*

3.1.1 Perceived Benefits of GenAI

GenAI offers significant advantages for students in their studies, with benefits afforded across the disciplines. Two of our students, from Dentistry and Computing and Information Systems, talked about GenAI as being an additional tool to support learning. This was expressed nicely by one of the students using the following analogy: "Just as keyboards and screens have not eliminated the use of physical paper for writing, GenAI will complement rather than replace traditional educational practices" (A7).

Time-saving and increases in efficiency were regularly reported benefits of GenAI tools amongst our students, with students across disciplines highlighting this as an advantage of using GenAI in their education, specifically through offering an opportunity for students to focus more on areas of particular interest (A6). GenAI tools' use in helping students explore knowledge and enhance their information discovery was identified as a benefit by several students; particularly the ability to assist with understanding large amounts of information and the potential to "understand or learn about any complex topic in the world in simple and conversational terms" (A2). Further advantages of GenAI for higher education identified by the students included providing 24/7 feedback and support, customisation of learning, assisting with revision, and fostering creativity. Our Computing student was the most positive regarding the use of GenAI in education, mentioning a large number of benefits and summarising with the belief that "the benefits of GenAI outweigh its negative impacts" (A7).

3.1.2 Perceived Risks of GenAI

Whilst most of our students weighed up the benefits of GenAI in education with its challenges, our English Literature student was much more pessimistic, and the only student not to identify any tangible benefits. This student believes that GenAI can encourage students to take short cuts, and be lazy, and that "it cheapens the work ethic of students and encourages them to rely on handouts" (A1). They suggest that students could even end up with an award that they don't deserve. This wariness, which is a possible reflection of the School's approach, is justified as follows: "The whole point of a literature degree is to develop your thinking, research, and writing skills, and GenAI compromises all of that. There is no point doing this degree, especially at postgraduate level, if you're not learning how to write better by yourself, and if you're not learning how to use and find credible sources for your argument" (A1).

There are, however, risks and challenges identified by most of the students, who generally showed an understanding of the limitations of using GenAI in their education. Over-reliance on GenAI was a key concern mentioned by most of our students, and missing out on acquiring key skills was a potential risk identified across the disciplines, particularly skills relating to critical thinking (A1, A4, A5) and creativity (A2 and A5). On the other hand, there is also an argument that GenAI tools can be seen to enhance creativity, a view shared by our Computing student. Ethical considerations were also clearly a concern, with most of our students mentioning these in some way. There was a particular focus on the impact on academic integrity, with one of our students commenting that those "who are not well aware of the ethical reasoning part of using AI might end up in a soup" (A2). The students were firm in putting the responsibility on the University to provide guidance and training for students and educators around the ethical aspects of GenAI use. Several students suggested that universities need to change their approach towards maintaining academic integrity: "I believe education institutions must rethink how to teach, test and expect learners to learn in a manner which minimises the effects of GenAI on academic integrity" (A4).

Perhaps surprisingly only one student, in International Business, mentioned the issues around "fictional or misinterpreted data" as a risk of GenAI use in education. This may suggest that concerns surrounding data reliability may not yet be fully understood or prioritised across all educational contexts.

3.1.3 GenAI and the University's Role

The challenges surrounding the roles and responsibilities of training in AI literacy and the provision of academic guidance and recommendation were voiced by our students. One student stated they don't believe it "is rational to expect all students to be aware and follow an ethical use of AI" and, similarly to the other students, highlighted the importance of clear guidance, as there is "nothing students hate more than uncertainty" (A4). Another student summarised the need for effective guidance: "The educational institute should decide and set up guidelines aligning to their values, ethics and integrity, because without shaping their narrative and marching into setting guidelines with vague ideas and vision will help no one and will further confuse everyone involved including students" (A6). In addition, there were also suggestions from our students that "academic staff also need to be trained appropriately in order to guide students" (A6).

Further student reflections highlighted concerns about the impact of the rapid development of GenAI tools on the University, which almost creates "an arms race between GenAI and educational institutes" (A4). Our Computing student is much more confident about how the institution can make use of this evolution in technology which "challenges teachers and educators to think creatively and innovate in ways that go beyond GenAI, enriching the learning experience" (A7). This student provided a very forward-thinking summary of the potential impact of GenAI on education: "This is a necessary and exciting force that propels our society's learning curve, ushering us into a new era … it is high time we discuss how to use GenAI, rather than debating whether to use it. Its integration is inevitable, and we cannot overlook the immense conveniences it offers" (A7).

3.2 *Theme 2: Impact of GenAI on Research*

The use of digital technologies to support the research process is well established. Examples of applications include reference managers, thematic analysis tools and online search engines. GenAI is the latest and potentially most powerful technology to transform how research is conducted. Despite the undoubted potential of GenAI to drive future research, our students were very aware of the challenges it presents to the research discipline including ethics, data protection, plagiarism and accuracy.

3.2.1 Perceived Benefits of GenAI to Research

It is clear from our research that GenAI is used extensively by students in their research activities. GenAI is regarded as a support tool which can accelerate the research process, provide insight into large volumes of data, act as the foundation for the generation of new research insights and ultimately create the research content itself. Interestingly, the further one moves down the list from a mere processor or interrogator of data to generating new content, the more sceptical and ethically-aware the students became. This scepticism is based on the quality of the research content and the ethical aspects around plagiarism and transparency, which is explored in the next section. In terms of specific examples, feedback from our students revealed that the potential benefits of GenAI to the research process include automation of research tasks, particularly those involving large amounts of data (A2). One of our students (A6) stated that GenAI "can save a lot of time" and another (A5) stated that it can act as "a short cut to finish/submit assignments". In both these cases however, the students added that this can negatively impact the research.

At a more strategic level, the students revealed that GenAI can potentially be utilised to not only enhance the efficiency of research but also to act proactively, as a form of co-authoring tool that goes beyond just processing data. Examples quoted for use in this way included "comprehending some part of the findings, finding new data for literature reviews and using its writing skills for outlining their research sections etc" (A2). As discussed regarding education, an additional benefit quoted by one student is the utility in GenAI to provide a useful starting point for understanding complex topics. In this respect, "GenAI could be used to gain a brief understanding of the topic research, improve grammar (similar to Grammarly) and help you create figures" (A4).

Interestingly, our dental student remarked that GenAI has the ability to improve accessibility to research, stating that GenAI "makes research more accessible to students from different backgrounds" (A4). This is supported by the findings that Chat GPT can assist in the development of lower-level cognitive skills exemplified by the lower levels of 'understand' and 'remember' in Bloom's taxonomy (Essien et al., 2024). However, while several students expressed the opinion that GenAI can introduce efficiencies to academic research, this was most often accompanied with reservations, and these are revealed in greater detail in the next section.

3.2.2 Perceived Risks of GenAI to Research

Many of the concerns expressed by students around the use of GenAI in research mirror those identified in higher education. In some respects, the concerns around the application of GenAI to research are a counterbalance to its benefits, namely around the erosion of critical thinking skills, irrelevant and inaccurate sources, ethical concerns and a general over-reliance on the tool. Our PhD student described using GenAI for research as a "…shortcut to finish/submit assignments without going through proper thought process". Furthermore, another student perspective stated that using GenAI excessively is a "wrong move", and that until GenAI models improve, uses should be limited to "…comprehending some parts of the findings, finding new data for lit reviews etc [or] using its writing skills for outlining research sections…" (A2).

A major concern of the students is that GenAI should not replace a strong understanding of the research topic. This was captured by our Computing student who stated, "from my experience, unless you know how to design a well-structured prompt... GenAI is likely to provide very generic answers" (A7). Students were also concerned about the risk inherent in AI-based searches leading to irrelevant sources and potential plagiarism. Particularly within English Literature, where frequently searched texts are promoted by LLMs over those which may be more relevant; a case of quantity over quality.

More than one student had concerns that the use of GenAI may encourage superficial research and uncritical thinking. Indeed, the importance of being able to defend why one chooses a specific text to support an argument was emphasised; by relying on GenAI, they posit that the researcher is not developing their critical thinking skills and is entering a "scholarly echo chamber" (A1). This view reveals a potential risk in over-reliance on GenAI by the student researcher; "It is natural for students to become reliant on the tool, but anyone who uses it frequently will realise that it cannot provide comprehensive assistance without some foundational knowledge of the subject" (A7). The concern is that students may lose the ability to think critically and instead rely solely on the output of GenAI.

Our students appreciated GenAI's capacity to process large amounts of data, helping to deliver insights that otherwise might take the human researcher considerably more time. However, there was also a recognition that there are considerable risks when utilising GenAI data which has no identifiable source, lacks transparency and can undermine the validity of the research. For example, there was awareness that the use of GenAI to generate content can potentially put the researcher at risk of academic misconduct. Our English Literature student stated that a peer who used GenAI to identify potential sources is "…putting herself at risk of plagiarism because GenAI isn't a source, but it may have picked a word order from a published scholar so it's a risky thing to do" (A1).

In the same context, our students expressed ethical concerns about data transparency and how this can undermine the quality and validity of research; "it saves a lot of time, however on the contrary, it can have a heavy impact on the transparency and truthfulness of the data and not to forget the ethics" (A7).

3.3 *Theme 3: Implications of GenAI for Future Careers*

The evolution of GenAI will continue to disrupt educational experiences, not only in how students learn, create and innovate but also how they will be expected to work following graduation. Our students offered unique outlooks on the impact GenAI could have on their future careers based on their respective disciplines.

3.3.1 Student Perspectives Across Academic Disciplines

The awareness of the need for upskilling was shared by all students, but particularly those in business and IT, reflecting an appreciation of current trends in employment and the direction of the future job market. Our International Business student aptly stated, "…the challenge here is how quickly can we develop the necessary skills to show our knowledge and adaptability" (A2).

Students felt GenAI could assist with routine tasks like blurb writing, report writing or legal writing. Our dental student similarly felt GenAI could assist with simple tasks like clinical notetaking. This sentiment, that GenAI "could take over and do a lot of (slightly) mundane human jobs" (A2) was shared by our students across multiple disciplines, highlighting GenAI as a tool with the ability to augment worker abilities. However, the implications of GenAI on diminished creative skill acquisition was highlighted as a key risk within education and research by our students (A2, A5); a concern which while rooted in an academic context, likely extends into the professional workplace. Furthermore, all students recognised the value and importance of human input, particularly within healthcare (A4, A5). Even within IT, the human element cannot be dismissed, with skills such as communication and explaining technical details to stakeholders remaining crucial in our students' eyes. Similarly, confidentiality within the healthcare field is of paramount importance, and our PhD student raised concerns over potential fraud stemming from inappropriate use of AI tools.

Ethical and legal challenges were raised by several students (A1, A5), namely highlighting issues relating to ownership or authorship of written work, copyright, accountability and author's rights with respect to AI-generated content in the workplace. Our English Literature student felt strongly that there is a need to protect "…the integrity of authors, whose work has been milked to death by AI" (A1). A further insight shared by our Human Resources Management student related to the fairness and equity around recruitment and automation. Despite the time-saving benefits, the increased use of AI resume screening tools means organisations need to question whether their process is "fair enough for the candidates applying? Is everyone getting an equal opportunity?" (A6).

4 Discussion: The Future and Good Practice

As we look to the future, the trajectory of GenAI in postgraduate education promises not only to redefine the tools and methods available for research and learning, but also to shape careers and disciplines. With GenAI models becoming increasingly sophisticated, the capacity for these tools to assist in both simple and complex tasks and foster new areas of academic inquiry is profound. The use of GenAI tools to assist with literature searching and knowledge acquisition are benefits which have potential to revolutionise education and improve learning outcomes (Yusuf et al., 2024). The student narratives reflect findings in the literature (Alam et al., 2023; Chan & Hu, 2023), whereby students are afforded increases in efficiency and time-saving, personalisation of learning, 24/7 feedback and support, with further benefits conferred to their revision and creativity. Perceptions of good practice in the use of GenAI and other AI tools will be shifting at an exponential rate to match the development of technology. However, certain key principles will remain, at least for the foreseeable future. In concluding this discussion, it is crucial to identify the good practices that will enable postgraduate students and educators to maximise the benefits of GenAI in their academic and professional journeys.

4.1 Human Input and Personalised Learning

GenAI will inevitably be able to do much of what humans do, and the act of plagiarism in higher education and research will likely remain a grey area as institutions continually play catch-up with technology, in a race to maintain academic integrity. Many of our students expressed concerns that misuse may undermine the value of their academic work (A1, A4, A5). Identifying whether GenAI has resulted in direct or accidental plagiarism is not a precise science, whether through digital AI detector software or when the output is reviewed by human experts (Gao et al., 2023). Researchers at all levels should therefore take great care when considering whether to integrate GenAI content into their work. Indeed, an overriding truth within this continually changing landscape may not be the definition of plagiarism, but the concept of the value that the individual brings to a piece of work.

Knowledge (which has become more freely available), and tools (whose scope in intellectual work has changed massively), are increasingly taking over the roles traditionally performed in 'human synthesis'. Human synthesis is the key to work having value to an educational institution or employer and avoiding plagiarism. If there was no human synthesis required, that job, course, or field of research would cease to exist. This is reflected in our student perceptions, that for your work to have value, human synthesis must play a significant role in organising and critiquing the knowledge, and effectively harnessing, but not over-relying on the tools (Chan & Hu, 2023). Indeed, several of our students (A2, A4, A6, A7) recognised the importance and value of prioritising the development of essential skills, specifically critical thinking and creativity.

If GenAI is capable of abstracting and critiquing knowledge, and providing insights, then human synthesis needs to shift to higher order tasks such as integrating, judging, reflecting, designing, and creating, relating again to Bloom's taxonomy (Newton et al., 2020). Therefore, as a tool, GenAI should be used to elevate the standard of work, and there is evidence that GenAI can promote creativity, critical thinking and problem-solving skills (Yusuf et al., 2024).

Personalised learning is the concept of taking learning and making it fit one's individual circumstances, learning preferences, speed, and interests to improve the efficiency of learning (Melzer, 2018). Typically, this has relied on time-consuming adaptations by educators for individual students, and more recently, the application of technology to offer choice and support, albeit simplistically or with great human input (Nerem et al., 2024). GenAI offers a means to streamline the process of adapting learning materials, with the ability to simplify explanations, re-order learning, recommend resources, and change the format of learning (see Fig. 2.1).

Multiple studies have looked at large data sets for predicting performance at university that have high accuracy and can retroactively predict the needs of students (Liles et al., 2018; Thiele et al., 2016). We now have the ability to do this in real-time. Moving forwards, GenAI tools may be used to analyse assessments, combined with entry grades and personal characteristics, to provide solutions to the individual such as what topics to focus on for study, and indeed how to study them.

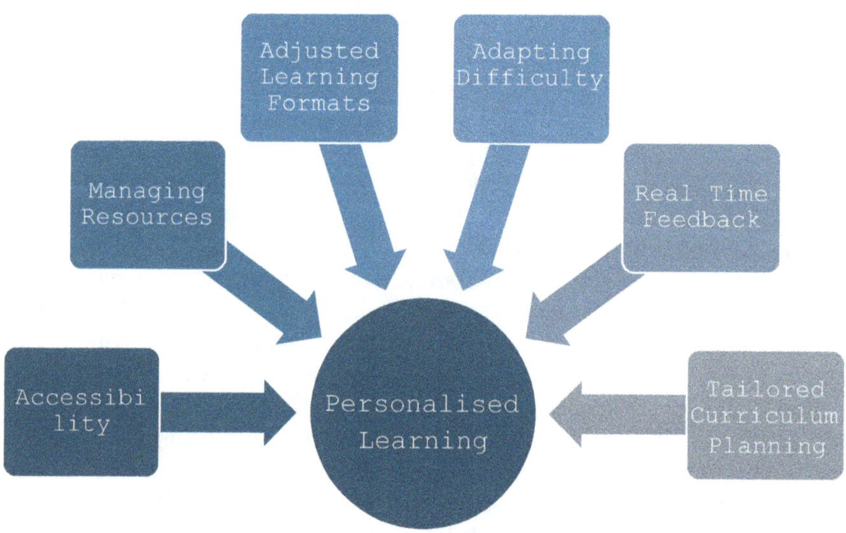

Fig. 2.1 Aspects of personalised learning with the use of GenAI

4.2 Research Concerns and Moving Forwards

The students interviewed for this chapter revealed a critical viewpoint on the utility of GenAI for academic research. On the one hand, they recognised the benefits of GenAI in introducing efficiencies and increasing accessibility to research; echoing findings from the literature such as the ability to generate abstracts (Gao et al., 2023), conduct literature reviews (Müller et al., 2022) and expedite academic publishing (Kitamura, 2023; van Dis et al., 2023). At a more complex level, GenAI may also be useful in summarising large numbers of published articles (Berg, 2023; Chan & Zhou, 2023), potentially operating at a meta level, identifying trends and gaps in research (Burger et al., 2023; Nguyen-Trung et al., 2023). There is also the perception that GenAI may improve accessibility in the face of language barriers. However, this can be contrasted with the view that in the case of second language learners, creating prompts can be challenging as they require a level of competence in written English, and over-reliance on GenAI can potentially hinder the development of necessary language skills (Warschauer et al., 2023). Students appear to view GenAI as a useful starting point in their research, whereby the student researcher can build on the initial insights and develop these further. This feature has also been described as a 'cognitive amplifier' by Essien et al. (2024), freeing the student researcher to engage in higher level thinking. Therefore, GenAI moves beyond its role as an expediter of repetitive tasks to a tool that can augment human capabilities (Butson & Spronken-Smith, 2024).

Despite the benefits afforded by GenAI, students highlighted the risks inherent in over-reliance on the LLMs. These include the erosion of critical thinking skills and the introduction of a range of ethical issues that can undermine the validity of the research. This concern is echoed in the literature, with Farrokhnia (2024) noting that ChatGPT lacks the capability to evaluate whether the data it generates is indeed accurate. This can lead to the fabrication of sources that are either inaccurate or do not exist at all. A further concern expressed was that GenAI is unable, due to its reliance on pattern recognition, to identify causal relationships (Bishop, 2021). This is particularly predominant in the humanities and social sciences which favour qualitative research methodologies (Butson & Spronken-Smith, 2024). Furthermore, the use of GenAI in research raises issues around data privacy and informed consent, challenges which researchers must grapple with leveraging GenAI's capabilities (Butson & Spronken-Smith, 2024).

Quite how student researchers and higher education can navigate these challenges is, at the time of writing, a work in progress, but the ability to think critically and express independent thought in research must be a key priority for the development of students (Golden, 2023). At a philosophical level, GenAI can be viewed as embodying technological determinism in which its effect on research is outside the control of regulatory bodies and the researchers themselves. At the other end of the continuum, advocates of social determinism propose that GenAI's place in research is mediated by human actions and decisions. According to UNESCO (2023a) the full potential of AI in research can only be realised if it is "rigorously reviewed and steered by a human-cantered approach". From the survey results, it is very possible that our students would advocate this approach.

4.3 Preparing for an AI-Enabled Workplace

There is clearly a need to anticipate and embrace new opportunities which could help to mitigate the negative effects and capitalise on the potential benefits of AI integration in all future careers. This outlook was shared by our Computing and Information Systems student who "looks forward to discovering which new roles I might pivot to in order to stay relevant in the evolving landscape" (A7).

This echoes employers' responses to the Amazon Web Services (2023) survey, who expect AI to dramatically impact specific industries, namely IT, Business operations, Sales and Marketing and Finance. Similarly, employers in legal sectors and human resources felt AI would be impactful and beneficial. Both employers and employees highlighted aspects such as automation, improved workflow, enhanced communication, learning new skills and boosts to innovation and creativity. Interestingly, increased creativity and innovation were the most reported benefits from the employer's viewpoint, whilst employees favoured benefits through automation of repetitive tasks.

Technical skills will become increasingly sought after as we move into the future (see Table 2.1), highlighting the clear need for AI literacy in the workplace to augment worker abilities (Zirar et al., 2023). All our students highlighted the need for a balance with human-centric skills (Brydon, 2019; Zirar et al., 2023) such as editing, critical thinking, and within healthcare, the requirement of clinician expertise and judgment alongside patient interactions (Davenport & Kalakota, 2019). While the application of AI within healthcare holds huge potential, misuse could have disastrous effects on patient health outcomes and trust in the profession without adequate regulation. Addressing the intersection where automation meets human judgement is key to ensuring ethical use and maintaining data security. The following competencies were identified by our students as key areas warranting focused development and upskilling for the effective management of GenAI in education, research and the workplace:

1. **Prompt Engineering**: Prompt engineering describes giving instructions to a GenAI tool and involves setting the context and expectations of the task. The more detailed the information provided, the more precise the output will be. As mentioned, our students (A2, A7) felt the ability to create succinct prompts that deliver the intended results is the key to maximising use of GenAI as a tool. Maloy and Gattupalli (2024) discuss the most important factors as: specificity, context, clear language, boundaries e.g. word count, iteration (refining the AI-generated response), and avoiding ambiguity.
2. **Interpretation**: GenAI tools are prone to making errors or displaying biases, and occasionally go on inexplicable tangents known as 'hallucinations'. It is

important that an effective user of the tool can spot and correct where the AI has made an error (A2). Those higher order skills that involve metacognition and procedural knowledge, and the analytical, evaluative, and creative tasks, are currently where human intervention is very necessary. It is these domains that one should focus on developing through education. Although this was only raised by our Business student as a risk within education, other studies clearly highlight awareness of the issues around the accuracy of both algorithms used and output (Amoozadeh et al., 2024; Chan & Hu, 2023; Yusuf et al., 2024).

3. **Data Literacy and Training**: On a surface level, GenAI draws information from a database, and orders it in the most likely outcome for the input it has received. What database that is, and the methods by which it orders information and interprets input, will change by AI tool and over time through updates. Students recognised the importance when relying on data produced elsewhere, that we understand the processes and potential biases by which AI-generated output is produced, particularly where this may lead to the spread of fraudulent information. For example, "false videos for political purposes" or the use of "fabricated data that requires critical thinking and additional verification before using or citing" (A5).

This insight will allow individuals to make key decisions when evaluating what is created. For this reason, almost all our students (A2, A4, A6, A7) clearly want to see universities providing training and advice on using GenAI in their education (Chan & Hu, 2023; Johnston et al., 2024), with the responsibility falling to universities to provide recommendations and guidelines which align with institutional "values, ethics and integrity" (A6) to maintain academic integrity (Kleebayoon & Wiwanitkit, 2023).

5 Conclusion

In this chapter we've explored the views of postgraduate students, who emphasised the importance of adapting higher education to the advancements of GenAI. They advocate for embedding a future-focused approach that embraces change and readies students for new AI-driven career opportunities. In recognition of the transformative potential of AI, students must approach GenAI with a balanced perspective and reflect on its impact on their future goals and careers. Echoing calls from UNESCO (2023b), higher education institutions and employers need to adopt a strategic approach which forefronts human-centric skills development with an AI-focus. It is imperative that students embrace the potential of GenAI, while understanding its limitations, to develop the skillset to navigate the ever-changing landscape of an AI-enhanced world. By establishing clear guidelines and ethical frameworks, higher education can adapt and evolve, harnessing GenAI's potential to enhance learning and better prepare students for the evolving job market. The gradual integration of AI tools, coupled with comprehensive AI literacy and ethics education, ensures that both students and staff can use these technologies responsibly and ethically.

Acknowledgements We would like to acknowledge and thank the QMUL student participants for completing our survey and consenting to share their views, some of whom requested to remain anonymous. These include: Student A1 (Kasandra Mavrigiannaki), Student A2 (Diksha Prasanna), Student A3 (Anonymous), Student A4 (Rawand Shado), Student A5 (Anonymous), Student A6 (Anonymous) and Student A7 (Khanh Tran).

References

Alam, F., Lim, M. A., & Zulkipli, I. N. (2023). Integrating AI in medical education: Embracing ethical usage and critical understanding [Opinion]. *Frontiers in Medicine, 10.* https://doi.org/10.3389/fmed.2023.1279707

Amazon Web Services. (2023). *Accelerating AI skills: Preparing the workforce for jobs of the future.* Retrieved July 29, 2024 from https://assets.aboutamazon.com/e1/a0/17842ee148e8af9d55d10d75a213/aws-accelerating-ai-skills-us-en.pdf

Amoozadeh, M., Daniels, D., Nam, D., Kumar, A., Chen, S., Hilton, M., Ragavan, S. S., & Alipour, M. A. (2024). Trust in Generative AI among students: An exploratory study. 67–73. https://doi.org/10.1145/3626252.3630842.

Attewell, S. (2023). *How will generative AI affect students and employment?* Luminate Prospects. Retrieved July 29, 2024 from https://luminate.prospects.ac.uk/how-will-generative-ai-affect-students-and-employment

Bearman, M., Ryan, J., & Ajjawi, R. (2023). Discourses of artificial intelligence in higher education: A critical literature review. *Higher Education, 86*(2), 369–385. https://doi.org/10.1007/s10734-022-00937-2

Berg, C. (2023). The case for generative AI in scholarly practice. *SSRN Electronic Journal.* https://doi.org/10.2139/ssrn.4407587

Bishop, J. M. (2021). Artificial intelligence is stupid and causal reasoning will not fix it [review]. *Frontiers in Psychology, 11.* https://doi.org/10.3389/fpsyg.2020.513474

Brydon, A. (2019). Why AI needs human input (and always will). *Forbes.* Retrieved 14.8.2024 from https://www.forbes.com/councils/forbestechcouncil/2019/10/30/why-ai-needs-human-input-and-always-will/

Burger, B., Kanbach, D. K., Kraus, S., Breier, M., & Corvello, V. (2023). On the use of AI-based tools like ChatGPT to support management research. *European Journal of Innovation Management, 26*(7), 233–241. https://doi.org/10.1108/EJIM-02-2023-0156

Butson, R., & Spronken-Smith, R. (2024). AI and its implications for research in higher education: A critical dialogue. *Higher Education Research & Development, 43*(3), 563–577. https://doi.org/10.1080/07294360.2023.2280200

Chan, C. K. Y., & Hu, W. (2023). Students' voices on generative AI: Perceptions, benefits, and challenges in higher education. *International Journal of Educational Technology in Higher Education, 20*(1), 43. https://doi.org/10.1186/s41239-023-00411-8

Chan, C. K. Y., & Zhou, W. (2023). Deconstructing student perceptions of generative AI (GenAI) through an expectancy value theory (EVT)-based instrument. ArXiv, abs/2305.01186. https://doi.org/10.48550/arXiv.2305.01186

Coffey, L. (2023). *Students outrunning faculty in AI use.* https://www.insidehighered.com/news/tech-innovation/artificial-intelligence/2023/10/31/most-students-outrunning-faculty-ai-use

Crompton, H., & Burke, D. (2023). Artificial intelligence in higher education: The state of the field. *International Journal of Educational Technology in Higher Education, 20*(1), 22. https://doi.org/10.1186/s41239-023-00392-8

D'Agostino, S. (2023). *AI raises complicated questions about authorship*. Inside Higher Education Retrieved August 14, 2024, from https://www.insidehighered.com/news/tech-innovation/2023/08/22/ai-raises-complicated-questions-about-authorship

Davenport, T., & Kalakota, R. (2019). The potential for artificial intelligence in healthcare. *Future Healthcare Journal, 6*(2), 94–98. https://doi.org/10.7861/futurehosp.6-2-94

Department for Education. (2023). *Generative artificial intelligence in education*. Departmental statement. Retrieved August 21, 2023, from https://assets.publishing.service.gov.uk/government/uploads/system/uploads/attachment_data/file/1146540/Generative_artificial_intelligence_in_education_.pdf

Dwivedi, Y. K., Kshetri, N., Hughes, L., Slade, E. L., Jeyaraj, A., Kar, A. K., Baabdullah, A. M., Koohang, A., Raghavan, V., Ahuja, M., Albanna, H., Albashrawi, M. A., Al-Busaidi, A. S., Balakrishnan, J., Barlette, Y., Basu, S., Bose, I., Brooks, L., Buhalis, D., et al. (2023). Opinion paper: "So what if ChatGPT wrote it?" Multidisciplinary perspectives on opportunities, challenges and implications of generative conversational AI for research, practice and policy. *International Journal of Information Management, 71*, 102642. https://doi.org/10.1016/j.ijinfomgt.2023.102642

Essien, A., Bukoye, O. T., O'Dea, X., & Kremantzis, M. (2024). The influence of AI text generators on critical thinking skills in UK business schools. *Studies in Higher Education, 49*(5), 865–882. https://doi.org/10.1080/03075079.2024.2316881

Farrokhnia, M., Banihashem, S. K., Noroozi, O., & Wals, A. (2024). A SWOT analysis of ChatGPT: Implications for educational practice and research. *Innovations in Education and Teaching International, 61*(3), 460–474. https://doi.org/10.1080/14703297.2023.2195846

Gao, C. A., Howard, F. M., Markov, N. S., Dyer, E. C., Ramesh, S., Luo, Y., & Pearson, A. T. (2023). Comparing scientific abstracts generated by ChatGPT to real abstracts with detectors and blinded human reviewers. *npj Digital Medicine, 6*(1), 75. https://doi.org/10.1038/s41746-023-00819-6

Golden, B. (2023). Enabling critical thinking development in higher education through the use of a structured planning tool. *Irish Educational Studies, 42*(4), 949–969. https://doi.org/10.1080/03323315.2023.2258497

Johnston, H., Wells, R. F., Shanks, E. M., Boey, T., & Parsons, B. N. (2024). Student perspectives on the use of generative artificial intelligence technologies in higher education. *International Journal for Educational Integrity, 20*(1), 2. https://doi.org/10.1007/s40979-024-00149-4

Joint Information Systems Committee (JISC). (2023). *A generative AI Primer*. Retrieved August 21, 2023, from https://nationalcentreforai.jiscinvolve.org/wp/2023/05/11/generative-ai-primer/

Kimbrough, K. (2024). *AI has arrived at work*. LinkedIn economic graph. Retrieved July 29, 2024 from https://economicgraph.linkedin.com/blog/ai-at-work-has-arrived

Kitamura, F. C. (2023). ChatGPT is shaping the future of medical writing but still requires human judgment. *Radiology, 307*(2), e230171. https://doi.org/10.1148/radiol.230171

Kleebayoon, A., & Wiwanitkit, V. (2023). Artificial intelligence, Chatbots, plagiarism and basic honesty: Comment. *Cellular and Molecular Bioengineering, 16*(2), 173–174. https://doi.org/10.1007/s12195-023-00759-x

Kramm, N., & McKenna, S. (2023). AI amplifies the tough question: What is higher education really for? *Teaching in Higher Education, 28*(8), 2173–2178. https://doi.org/10.1080/13562517.2023.2263839

Kuadey, N. A., Ankora, C., Tahiru, F., Bensah, L., Agbesi, C. C. M., & Bolatimi, S. O. (2023). Using machine learning algorithms to examine the impact of technostress creators on student learning burnout and perceived academic performance. *International Journal of Information Technology., 16*, 2467–2482. https://doi.org/10.1007/s41870-023-01655-3

Liles, J., Vuk, J., & Tariq, S. (2018). Study habits of medical students: An analysis of which study habits Most contribute to success in the preclinical years [version 1]. *MedEdPublish, 7*(61). https://doi.org/10.15694/mep.2018.0000061.1

Malik, T., Hughes, L., Dwivedi, Y. K., & Dettmer, S. (2023). *Exploring the transformative impact of generative AI on higher education*. New sustainable horizons in artificial intelligence and digital solutions (pp. 69–77). Cham.

Maloy, R. W., & Gattupalli, S. (2024). *Prompt literacy*. EdTechnica: The Open Encyclopedia of Educational Technology. https://doi.org/10.59668/371.14442

Melzer, P. (2018). *A conceptual framework for personalised learning: Influence factors, design, and support potentials*. Springer Fachmedien Wiesbaden.

Müller, H., Pachnanda, S., Pahl, F., & Rosenqvist, C. (2022). The application of artificial intelligence on different types of literature reviews – A comparative study. *International Conference on Applied Artificial Intelligence (ICAPAI), 2022*, 1–7. https://doi.org/10.1109/ICAPAI55158.2022.9801564

Nerem, T. S., Hellesnes, S. F., Inal, Y., & Monllaó, C. V. (2024). *The use of artificial intelligence for personalized learning: Teacher perspective*. Retrieved July 30, 2024, from https://www.thinkmind.org/library/ACHI/ACHI_2024/achi_2024_3_110_20056.html

Newton, P. M., Da Silva, A., & Peters, L. G. (2020). A pragmatic master list of action verbs for bloom's taxonomy [brief research report]. *Frontiers in Education, 5*. https://doi.org/10.3389/feduc.2020.00107

Nguyen-Trung, K., Saeri, A. K., & Kaufman, S. (2023). Applying ChatGPT and AI-powered tools to accelerate evidence reviews. *OSF Preprints*. https://doi.org/10.31219/osf.io/pcrqf

Quality Assurance Agency. (2023a). *Maintaining quality and standards in the ChatGPT era: QAA advice on the opportunities and challenges posed by Generative Artificial Intelligence*. Retrieved April 30, 2024, from https://www.qaa.ac.uk/docs/qaa/members/maintaining-quality-and-standards-in-the-chatgpt-era.pdf?sfvrsn=2408aa81_10

Quality Assurance Agency. (2023b). *QAA briefs members on artificial intelligence threat to academic integrity*. Retrieved November 21, 2023, from https://www.qaa.ac.uk/news-events/news/qaa-briefs-members-on-artificial-intelligence-threat-to-academic-integrity#

The Russell Group of Universities. (2023). *Russell Group principles on the use of generative AI tools in education*. Retrieved August 21, 2023, from https://russellgroup.ac.uk/news/new-principles-on-use-of-ai-in-education/

Thiele, T., Pope, D., Singleton, A., & Stanistreet, D. (2016). Role of students' context in predicting academic performance at a medical school: A retrospective cohort study. *BMJ Open, 6*(3), e010169. https://doi.org/10.1136/bmjopen-2015-010169

Thorpe, J. (2023). *The real risk of generative AI is a crisis of knowledge*. WONKHE. Retrieved March 01, 2024, from https://wonkhe.com/blogs/the-real-risk-of-generative-ai-is-a-crisis-of-knowledge/

UNESCO. (2023a). *Guidance for generative AI in education and research*. Retrieved August 10, 2024 from https://www.unesco.org/en/articles/guidance-generative-ai-education-and-research

UNESCO. (2023b). *Oxford and Cambridge ban ChatGPT over plagiarism fears*. International Institute for Educational Planning. Retrieved April 30, 2024, from https://etico.iiep.unesco.org/en/oxford-and-cambridge-ban-chatgpt-over-plagiarism-fears

van Dis, E. A. M., Bollen, J., Zuidema, W., van Rooij, R., & Bockting, C. L. (2023). ChatGPT: Five priorities for research. *Nature, 614*(7947), 224–226. https://doi.org/10.1038/d41586-023-00288-7

Warschauer, M., Tseng, W., Soobin, Y., Webster, T. J., Jacob, S., Du, Q., & Tate, T. (2023). The affordances and contradictions of AI-generated text for writers of English as a second or foreign language. *Journal of Second Language Writing, 62*, 101071–101071. https://doi.org/10.1016/j.jslw.2023.101071

World Economic Forum. (2023). *The future of jobs report 2023*. Retrieved July 29, 2024 from https://www.weforum.org/publications/the-future-of-jobs-report-2023/

Yusuf, A., Pervin, N., & Román-González, M. (2024). Generative AI and the future of higher education: A threat to academic integrity or reformation? Evidence from multicultural perspectives. *International Journal of Educational Technology in Higher Education, 21*(1), 21. https://doi.org/10.1186/s41239-024-00453-6

Zahidi, S. (2020). *The jobs of tomorrow*. International Monetary Fund. Retrieved July 29, 2024 from https://www.imf.org/en/Publications/fandd/issues/2020/12/WEF-future-of-jobs-report-2020-zahidi

Zirar, A., Ali, S. I., & Islam, N. (2023). Worker and workplace artificial intelligence (AI) coexistence: Emerging themes and research agenda. *Technovation, 124*, 102747. https://doi.org/10.1016/j.technovation.2023.102747

Chapter 3
AI Tools and Technologies for Academic Research

Attila Szendi, Dániel Kuttor, and Zsolt Pál

Abstract This chapter explores the landscape of Artificial Intelligence (AI) tools and technologies specifically suited for academic research, emphasizing the role of locally run Generative AI (GenAI) applications. By examining the structure of data management technologies and AI software components, the chapter illustrates how researchers can leverage data-driven AI systems to build knowledge bases from scientific data. It highlights the value of Retrieval-Augmented Generation (RAG) systems for processing and referencing local documents, underscoring their relevance to the knowledge-intensive nature of research. Additionally, the chapter discusses open-source software (OSS) for ensuring transparency and repeatability, and reviews essential hardware configurations needed for efficient local AI deployment. This work aims to equip researchers with the understanding required to select AI tools that enhance their research workflows, offering insights into both foundational principles and practical applications of AI in academic settings.

1 Introduction

The Artificial Intelligence (AI) technology is a dynamic and complex field that is at the centre of many research interests and in the major focus of R&D activities by technology-intensive companies. A number of online resources, such as Wikipedia or various encyclopaedias, dictionaries, software collection, offer a way to navigate among the available services. However, there is much less information available on locally run Generative AI applications, which are particularly important for research

D. Kuttor (✉) · Z. Pál
Faculty of Economics, University of Miskolc, Miskolc, Hungary
e-mail: daniel.kuttor@uni-miskolc.hu

A. Szendi
Library, Archives, Museum, University of Miskolc, Miskolc, Hungary

X. Zhou, H. Al-Samarraie (eds.), *Institutional guide to using AI for research*,
Advances in Artificial Intelligence in Education 2,
https://doi.org/10.1007/978-3-031-94809-1_3

as they provide a deeper insight into the functioning of Generative AI systems. We aim to describe this knowledge in the chapter.

It is important to note at the beginning that GenAI aims to create new content using pre-learned data sets. Thus, the data asset, which is a scientific, verified data repository, is proved to be more important than the role of Large Language Models (LLMs). Based on these processed, verified scientific data sources, we can talk about scientific research supported by AI tools and technologies (Russell & Norvig, 2021).

2 Components of Data Management Technologies: Structure and Main Attributes of AI Software

Our work aims to provide an overview of the most used AI tools, and the technologies needed to understand how to use them. Our aim is to enable the reader to select and test the most appropriate applications for research. Understanding how the latest AI technologies work is essential, at least at the level of the basic principles. In this context, we emphasise the need to speak about AI software as a system of data and algorithms. With this knowledge, we can select applications whose choices are no longer necessarily about the specific software, but about the principles that the software implements (Li et al., 2024) (Fig. 3.1).

Data-driven AI

The most modern AI systems belong to the category of data-driven AI systems, the most important feature of which is that they use different algorithms (Convolution Neural Network <CNN>, Recurrent Neural Network <RNN>, transformer) to build a knowledge base from the available data, which generates answers to the user's free-word questions and instructions based on statistical probability.

Generative AI

Tools >>>> Technology >>>> Result

AI software: data-driven	**Backends:** API for LLM	**Source of Data:** unstructured, structured		**Generated Data:**
Target: Natural language Processing (NLP) and multimodal tasks Job: Prompt writing, Autonomous agents Developing: Framework, libraries	RAG technology for local docs: vector database Hardware: CPU, GPU	*Datasets for LLM* Prompt text in long-context modelling Local + web Docs for knowledge-intensive domains Computer Vision	*Large Language Model (LLM)* Test only Multimodal Domain-specific Pretrained Quantized GGUF file-format	Text paraphrased Text: with cited document Image Audio, Voice Video 3D *(10-20% hallucination)*

Principles of research	1. Open-source software vs. proprietary software 2. Open data vs. Closed data 3. Local Desktop APP, self-hosted vs. cloud provider

Fig. 3.1 Structure and components of the GenAI. Source: authors' compilation

Knowledge-Intensive Use

AI can handle two main data inputs: LLMs based on the Transformer architecture and Retrieval-Augmented Generation search. LLMs work by exploiting deep learning techniques and huge amounts of text data. The RAG technology facilitates the processing of local documents in cases where the necessary documents were not available when the LLM was trained, or where frequently changing data is not practical to handle by costly relearning of LLMs. Since scientific research is a knowledge-intensive activity, processing local documents is a priority for the application of AI. Thus, when describing the tools, it is important to consider the ability of the application to handle local document-based technologies, such as RAG or similar solutions (Wei et al., 2022).

Hallucination-Confabulation

One of the biggest challenges in AI research is to filter the hallucination or confabulation (Latin: confabulatio, meaning "weaving a tale"). AI algorithms may generate outputs that are not based on trained data, resulting in findings that are incorrect. To reduce hallucinations, techniques that link the LLM response to reliable documents are being tried. This approach is known as the RAG technique and there are further, improved versions of it (Béchard & Ayala, 2024).

Open-Source Software (OSS)

Another important element of selection is how transparent and documented the operation of the AI application is at the LLM level. The repeatability of scientific research is based on the ability of another researcher to perform the same research using the same tools, but possibly with different settings. This can only be done with open-source software (OSS) and, most importantly, with an LLM (open LLM) trained on open datasets. Open-source LLM datasets are public data such as Common Crawl (Source: https://commoncrawl.org/impact) or RedPajama-V2 (Source and further information: GitHub[1]).

A special class of local data is research data (measurements, data collections), which can be sensitive data (copyright, personal data or other research interest). Processing these data with AI applications running in the cloud (cloud, API) is limited, usually with expensive subscriptions. The justification for a cloud-based service can be justified if it is used to perform a well-defined task with frequently varying, possibly extremely different hardware requirements. The obvious technology is on-premises LLM and its applications. This is the next aspect that we will keep in mind in our presentation. Open LLMs provide transparency, auditability and flexibility for research.

Prompts

AI applications can be purpose-built (Automatic Speech Recognition <ASR>, Handwritten Text Recognition <HTR>, Text-to-Speech <TTS>, etc.), but they can also be assistants integrated into other IT programmes. However, the greatest functionality is provided by individual AI applications. Their primary use is via a chat interface, but this interface is more like a command line (prompt) in IT

[1] https://github.com/togethercomputer/RedPajama-Data

terminals. Using the latter in this way offers more opportunities to exploit the capabilities of AI but also requires greater expertise. Non-expert users can be assisted by pre-made prompt templates. Two important techniques are the Chain-of-Thought (CoT) and the Emotion Prompt. Learning prompt techniques is one of the most useful skills for using AI.

Self-Hosted Solutions
Applications running locally can be either desktop software or multi-user systems that provide information through LLMs and local documents running on the server. Desktop applications usually include all the pre-built software dependencies needed for operating AI, such as downloadable LLMs, the ability to process local documents, and the frameworks and libraries (e.g. PyTorch). In contrast, server applications may include additional components that can be installed as a dependency of the main software but are optionally available for manual installation.

LLM
At the heart of Generative AI is data, including LLM, which has many variations due to the different settings of the source dataset and the transformer. A catalogue of the most used LLMs is usually included in desktop applications. Choose among them is not an easy task, and it is worth testing several language models. The production of language models trained for specific tasks requires a high level of expertise and an expensive process (Gao et al. 2020). Also, open LLMs are updated relatively frequently, which can be reported during desktop software updates. The most comprehensive database of models available on the Internet is Huggingface (Source: https://huggingface.co/models), and a list of models associated with the Ollama software can be usefully read and downloaded using the Ollama software (Source: https://ollama.com/library). LLMs that available online are a great help for testing (https://lmarena.ai/).

Data Source
Another key aspect of local AI applications besides LLM is the existence of local documents. There are two main problems: the analogue medium and the legal options. Digitisation is one of the most important scientific data assets building opportunities. Information that seems outdated and hidden in old literature can be reintroduced into the research process. It is worth exploring efficient hardware and software for digitisation (page-fed scanners, open-source Optical Character Recognition <OCR>, etc.). However, even more problematic is the need to respect copyright constraints when creating LLMs and collecting local documents. Non-profit scientific research has possibilities which are provided by legislators in EU and Anglo-Saxon common law countries. EU legislation in this area allows public collections and non-profit research institutions (universities) to use copyrighted works for the express purpose of text and data mining. This is a huge opportunity, knowing that the most important IT companies in AI currently suffer from a lack of quality, selected data. The data assets of research institutions can be transformed into a whole new semantically processed data asset through AI technology.

Hardware Parameters

In addition to data and software, the right physical framework, the hardware is an important element for local AI to work. The extremely high hardware costs requiring to train AI language models (LLMs) and the data centres of large cloud providers serving millions of users are not authoritative. Systems with thousands of GPUs (Graphics Processing Units) have been designed to handle workloads which relatively little number of research need.

In fact, AI applications running locally can be run at speeds sufficient for personal research use on a personal computer with an AVX2-compatible processor, 16 GB of RAM and 500 GB of SSD without a GPU, for example, using the llama.cpp (Source and further information: GitHub[2]) software library for quantizing LLMs. Quantization is a technique used in machine learning to reduce the computational and memory requirements of models.

For server-based hardware, a high-performance workstation with a 16–24 GB GPU card, 64–128 GB of RAM and a CPU with AVX2 or AVX512 instruction set can serve dozens of users simultaneously. GPUs compatible with NVIDIA's NVLink technology (nvlink-bridges) can be used to link up to two GPU cards, so your computer can see 2 × 20 = 40 GB or 2 × 24 = 48 GB of memory (Source: https://www.nvidia.com/en-us/design-visualization/nvlink-bridges/).

3 Functional Properties of AI

General Features of AI Information Processing

This section is presented according to the principles of AI applications in scientific research. These are, as we have already shown transparency, traceability and auditability. State-of-the-art AI is data-driven, so data is at its core. Scientific research also focuses on scientific literature and research data (Buolamwini 2023). Therefore, only AI software that is based on open, verifiable data is suitable for conducting scientific research. The basis for data-driven AI is LLM, which uses large amounts of unstructured and structured data. Knowing the data sources and datasets of LLM is essential to know what the results of Generative AI can be traced back to. However, the LLM of data-driven AI cannot contain all the data at the time of its training, as scientific research is constantly enriched with growing literature and research data. (Zhuang et al., 2024) The same problem arises with AI-driven processing of web content: it is not possible to create an LLM from documents on the web daily. Moreover, LLM does not store documents with references to specific factual claims. This is why complementary technologies (e.g. RAG and its enhancements) have been developed to enable document referencing and to handle frequent document changes with more resource-efficient technologies.

[2] https://github.com/ggerganov/llama.cpp

Adding local documents to the knowledge base of the AI is a crucial function. Accordingly, only software that offers this possibility will be described below (unless the exclusive existence of other specific features justifies the inclusion of additional software).

It is always important to emphasise that the application of AI in research can only be successful with LLM trained on a selection of high-quality data and additional scientifically valuable data. In practice, this can be most effectively achieved through the development of an institution-wide scientific data estate, which may involve the involvement of information science (library, archives, museum, etc.) professionals independent of individual researchers.

Data-driven AI technology is primarily implemented using transformer-based techniques to capture the semantic meaning of unstructured data. By employing sentence, word, and syllable embeddings, these AI systems are able to interpret and process content. The integration of retrieval-augmented generation (RAG) technology further enhances AI's ability to identify the sources and relevant text fragments used for responses, ensuring verification and traceability. Additionally, the core capabilities of LLMs include extraction, summarisation, translation, and answer generation, all of which are traditionally classified under NLP (Natural Language Processing). AI software can be categorised into goal-oriented (e.g., speech recognition), integrated (e.g., ASR, TTT, TTS), and autonomous AI agents. (Edge et al., 2024).

Categories of Generative AI Use

AI offers a wide range of possibilities for processing both unstructured and structured text data, as well as multimedia data (image, audio, video, 3D) in multimodal LLMs. Below are some grouped examples, some of which are suitable for general, everyday use.

- **Data creation:** idea generation, image generation, social media management, knowledge base generation.
- **Editing and transformation:** summarising, paraphrasing, explaining, language translation.
- **Programming:** code writing, code transformation, bug fixing, code explanation.

When considering the practical application of LLMs, further information processing tasks can be identified from the Huggingface database (Source: https://huggingface.co/tasks):

- **Natural Language Processing (NLP):** Text Generation, Question Answering, Sentence Similarity, Translation, Summarisation.
- **Audio:** Automatic Speech Recognition, Text-to-Speech.
- **Computer Vision:** Image Classification, Object Detection, Text-to-Image, Image Segmentation, Image-to-Text.
- **Multimodal:** Visual Question Answering.

Classifications of AI-Driven Information Processing Technologies

AI-driven technologies process and transform various inputs, including text, speech, images, video and sound, into new formats for diverse applications. The text-based tools cover the followings:

- Text-to-Text (TTT) for translation and content generation, Text-to-Speech (TTS) for accessibility, and Text-to-Code (TTC) for programming automation.
- Speech technologies like Speech-to-Text (STT) and Speech-to-Speech (S2S) support transcription and multilingual conversations.
- Image-focused tools include Text-to-Image (TTI), Image-to-Text (ITT), and Image-to-Image (ITI) for creative content and data extraction.
- Video processing is enabled through Text-to-Video (TTV), Video-to-Text (VTT), and Image-to-Video (ITV) for generating and summarizing media.
- Additionally, Code-to-Text (CTT) and Sound-to-Text (SnTT) facilitate code documentation and audio analysis, showcasing AI's versatility in transforming information across input-output relationships.

Integrated ASR-TTT-TTS Applications

The latest software solutions combine computer speech recognition, translation and text reading technologies. This enables live conversations between people speaking different languages.

- **Babelfish.ai** is a real-time live speech translator and translator application that runs locally in a browser. Everything happens in the browser using Transformers.js and ONNX Runtime Web. For translation, it uses Meta No Language Left Behind-200, which supports 200 languages. (Source and further information: GitHub[3])
- **Speech To Speech** is an open-source, modular GPT-4 based solution that implements speech-to-speech pipelines. It supports automatic speech recognition, Speech-to-Text (STT) using OpenAI Whisper, and Speech-to-Speech (STS) using Parler-TTS. Any Language Model is available on Huggingface can be applied. (Source and further information: GitHub[4])
- **Mini-Omni** is an open-source multimodal large language model that can listen and speak in real time. It has real-time speech input and streaming audio output with talk-talk capabilities. (Source and further information: GitHub[5])
- The **Speech Note** application can convert speech to text (ASR), text-to-speech (TTS) and machine translation (MT) in offline mode. It uses several different processing engines, such as Whisper.cpp ASR, PiperTTS, and the Bergamot translator, which uses the same translation engine as Firefox. (Source and further information: GitHub[6])

---

[3] https://github.com/alMubarmij/Babelfish.ailfish.ai

[4] https://github.com/huggingface/speech-to-speech

[5] https://github.com/gpt-omni/mini-omni

[6] https://github.com/mkiol/dsnote

Self-Hosted AI Applications

In addition to the importance of data transparency in research, it is essential to consider the interaction between the software environment running AI and LLMs (large language models). Software architecture plays a key role in shaping the performance and capabilities of LLMs. The transparency, sustainability and flexibility needed to achieve research goals can only be ensured through locally hosted open-source applications. Consequently, these two criteria—transparency and openness—become paramount in the analysis of AI applications, justifying the emphasis on the local availability of open LLMs discussed earlier. Furthermore, all components of the AI software architecture must be local, whether they are located on the user's computer or on the institution's server. In addition, it is essential to streamline the hardware requirements of the software to allow AI applications to run efficiently without special hardware components such as GPU cards. The hardware requirements have been discussed in detail in the introductory section.

The CPUs of computers manufactured after 2020 that support AVX2 instructions are well-suited to perform AI tasks. Expanding system memory and back-up storage capacity can eliminate the further limits of operation and hurry up the processes. Although some specific software may have different hardware requirements, in these cases 16–24 GB GPU cards represent the highest cost. It is important to note that for any configuration, a GPU card higher than 16 GB will help running of Generative AI but it is not considered as a critical requirement. The 7B models require at least 8 GB of RAM to run, while the 13B models require at least 16 GB.

A key aspect of running the software described is the ease of deployment. Most of the open-source software developed for this research is written in Python and generally uses a command-line deployment mechanism that can be easily used by a researcher with IT skills. Many of these programs also have a GUI, so the installation process is not a major challenge at all. A summary of software with a graphical installer will be given special attention to emphasise that researcher-run AI applications are not only accessible to IT professionals.

Python-based software categories deserve special attention from a research perspective, as most of the advanced AI systems and autonomous agents can be deployed and specified almost exclusively using this language.

A key element of the AI software architecture is the software component that handles the LLM. In addition to its operating mechanism, it has several hardware-related features. LLMs are typically 4–8 GB files, which are not practical to store separately per application. To this end, there are local LLM API servers to which applications can connect. For instance, the Ollama local LLM API server has a graphical installation and user interface and can be run without GPU support. The llama.cpp started out as a CPU-only solution but now aims to support all available computing devices and environments, including porting to the Android platform.

The importance of software with a graphical installer and user interface makes it worth collecting and highlighting them (Table 3.1).

Table 3.1 AI applications with Graphical User Interface and installer

AI Format (platform)	GUI and installer
Desktop	• GPT4all: supports 24 text-only models • LM Studio: supports 26 models, also supports Llava's Vison-enabled model • pyGPT: OpenAI API focused application demonstrating full functionality, but also supports Ollama local API server • Alpaca, about 140 models available in the Ollama library • TransformerLab: professional application also supporting Llava's multimodal model
Copilot applications	Web browser-based Page Assist and Brave Leo AI, Reor note-taking application with Ollama local LLM API server
Mobile	ChatterUI, Maid, MLCchat, Rtranslator, SherpaTTSPiper, NextGenKaldiASR
Integrated ASR-TTT-TTS	SpeechNote

Source: authors' compilation

Desktop AI Applications: The Importance of GUI

Due to the complexity of self-hosted AI, one of the most important aspects for the end client is the easy operation of the user interface. The more advanced customer is looking for rich functionality, either through the configuration options or by switching to another application.

GPT4all is a desktop AI application. The interface is transparent for the novice user, with the main functions on the landing screen, and the settings parameters allowing more advanced users to further fine-tune the application. The starting page contains buttons for the three most important functions: chat, local documents, models. It only works with text models. Hardware requirements are as described in the introductory section. (Source and further information: GitHub[7])

The LM Studio AI desktop application allows you to discover, download and run local LLMs. The application provides comprehensive information on the characteristics of locally executable models and provides direct access to online information sources such as the Huggingface platform. It is also able to work with multimodal LLMs. To ensure smooth operation, a local API server (Lmstudio engine) and an executable library (llama.cpp) are built in. The hardware requirements meet those described in the introductory section. (Source and further information: GitHub[8])

PyGPT AI application offers a mixed toolkit for local and remote models. It is highly recommended for advanced users. It supports a wide range of models: both online APIs (GPT-4, GPT-4 Vision, GPT-3.5, Gemini, Claude, DALL-E) and the Ollama local API (Llama 3). It allows embedding local document processing: Langchain, Llama-index libraries. In addition to chat, the PyGPT AI application supports multimodal features with a strong focus on computer vision, voice control,

[7] https://github.com/nomic-ai/gpt4all

[8] https://lmstudio.ai/

image generation (DALL-E 3) and analysis (GPT-4 Vision), file upload and download, speech synthesis and recognition (OpenAI Whisper). Agents: code and command execution, access to the web, assistants. Advanced features: chat memory, presets, plug-ins. (Source and further information: GitHub[9])

Alpaca is an Ollama client where you can use several models in the same conversation. It offers the most modeling capabilities of any graphical interface AI application (https://ollama.com/library). Features: image recognition with multimodal models, attaching text from a web page to a prompt, and processing local documents (txt, pdf). (Source and further information: GitHub[10])

Ollama is an open-source framework designed to facilitate the deployment of large language models in local environments. Applications using Ollama include Page Assist, PyGPT, Alpaca. Allows the use of about 140 maintained LLMs. (Sources: https://ollama.com/library)

Ollama App is a modern and user-friendly client for Ollama. The users can enjoy the maximum benefits of the software while maintaining the security and privacy of their data. This can be particularly important for applications that handle sensitive or personal data. (Source and further information: GitHub[11])

AI Assisted Software

In the case of AI-enabled software, we need to talk about two main components: the host application and the application that provides the AI service—hosted locally, according to our principles. The application developer must support AI at a plugin level at least.

The AI and the Web Browsing, Web Searching

Brave Browser is able to integrate local AI into the browser. Brave supports the "Bring Your Own Model (BYOM)" approach to AI functionality built into the Brave Leo browser: users have the ability to run local AI models using the Ollama LLM API server (Source and further information: GitHub[12]) with a graphical interface installer (Source: https://brave.com/blog/byom-nightly).

The Page Assist is developed for Chrome, as an open-source extension that provides a sidebar and web interface for interacting with Artificial Intelligence models on the local Ollama API server (Source and further information: GitHub). Users can summarize web pages and ask questions about content and get answers enhanced by web searches (Source and further information: GitHub[13])

The Most Advanced AI Applications: Systems of Agents

For AI applications to work successfully, by their design-concepts, multiple libraries or frameworks are necessary to be installed and operated. However, there is a need to run multiple agents to add action possibilities (executable instructions) to

---

[9] https://github.com/szczyglis-dev/py-gpt

[10] https://github.com/Jeffser/Alpaca

[11] https://github.com/JHubi1/ollama-app, https://github.com/ollama/ollama

[12] https://github.com/ollama/ollama

[13] https://github.com/n4ze3m/page-assist

the closed world of LLMs and thus to obtain new information. Multi-agent systems are able to solve such problems that cannot be fixed by a single agent. Agents generate their own chains of invocation after having formulated their goals: for tools, services, etc.

Through the Mixture-of-Agents (MoA) approach, the combination of different agents enhances the performance of large language models (Wang et al., 2024). With the increase in the number of LLMs, there are exciting opportunities to leverage the collective expertise of multiple LLMs, which can be implemented under the Mixture-of-Experts (MoE) concept. To this we propose a new approach that leverages the collective strengths of multiple LLMs within the MoE framework (Jiang et al., 2024). Each agent uses the outputs of the agents in the previous layer as auxiliary information to generate its response. (Source and further information: GitHub[14])

Research Agents

An available list of AI agents called "Awesome Agents" can be found in a description of open-source AI agents. (Source and further information: GitHub[15])

LLocalSearch is a wrapper (aggregator with codes) around the LLM running locally, allowing them to choose from a range of tools. These tools allow them to search the web for up-to-date information on specific issues. This process is recursive, which means that the running LLM is free to choose devices (even multiple times) based on information received from the user and other device calls. (Source and further information: GitHub[16])

The Gorilla empowers language models to leverage tools through API invocation. Upon receiving a natural language query, Gorilla identifies and suggests the most semantically and syntactically appropriate API for the task at hand. Accompanying this functionality is APIBench, which boasts the most extensive, well-curated collection of APIs, designed to be intuitive and easily accessible for users. (Source and further information: GitHub[17])

Multi-user GenAI Apps

Locally hosted genAI applications are not necessarily limited to separated, individual desktop versions. In some cases, it may be worthwhile to make a computer equipped with a GPU available to multiple users over a network and run the required AI application on it. This allows more efficient use of resources and avoids unnecessary copying of documents.

The Open WebUI (formally Ollama WebUI) allows the use of models that support multimodal interactions (e.g. Llava). Voice input support is possible: which means direct live verbal conversation with the models (GPT4-o). Progressive Web

[14] https://github.com/togethercomputer/moa

[15] https://github.com/kyrolabs/awesome-agents

[16] https://github.com/nilsherzig/LLocalSearch

[17] https://github.com/ShishirPatil/gorilla

Application (PWA) is available for mobile devices. (Source and further information: GitHub[18])

Onyx (formerly Danswer) is the AI application that connects to a group's (corporate, cultural, non-profit, academic) documents and applications, serving as a single search platform on all common work tools such as Slack, Google Drive, Confluence, etc. (Source and further information: GitHub[19])

BionicGPT serves as an on-premises alternative to ChatGPT, delivering the benefits of Generative AI while ensuring stringent data confidentiality. It can operate seamlessly on your laptop or scale up for data centre deployment (Source and further information: GitHub[20]).

The Harbor allows you to run LLM backends, APIs, frontends and services with a single command. Harbor is a compact LLM toolkit that implements integrated execution of LLMs and additional services. It consists of a CLI (command-line interface) and a companion application that implements easy management and execution of AI services (Source and further information: GitHub[21]).

Applications of Finetuning: Training

LLMs are monolithic large files, which are resource-intensive and costly to modify, a process called training. During training, the Transformer algorithm creates semantic embeddings from the available dataset. However, the resulting LLM may contain errors and mistakes from the dataset. This is compounded by the hallucination of Generative AI, which generates plausible but in fact false information about facts that are not in the dataset. To remedy this, fine-tuning procedures have been developed, which can be performed through GUI applications. Although the resource requirements of these applications do not reach the requirements of LLM training, they need at least a 24 GB GPU. The process itself requires more expertise, but there is scope for their use in a research project.

The Text Generation Web UI is built on the Gradio platform, specifically for LLMs. It is a feature-rich application that includes a simple LoRA (Low-Rank Adaption) fine-tuning tool that allows customization of models based on user input (Biderman et al., 2024). A special feature is the support for various additional extensions (Source and further information: GitHub[22]).

The Transformer Lab is an application that allows anyone to experiment with large language models on own computer (PC). Users can interact with the models, train and fine-tune them (Finetune, Train) and evaluate the performance of LLMs. The creation of datasets for training is also supported (Build Datasets for Training, Connect to Dataset Store). For best performance (e.g. when using RTX3090 or

[18] https://github.com/open-webui/open-webui

[19] https://github.com/onyx-dot-app/onyx

[20] https://github.com/bionic-gpt/bionic-gpt

[21] https://github.com/av/harbor

[22] https://github.com/oobabooga/text-generation-webui

RTX4090), VRAM GPUs with at least 24 GB RAM are recommended. The application is also compatible with plugins (Source and further information: GitHub[23]).

Krtansformers is a flexible framework to explore state-of-the-art LLM inference optimizations. It provides an innovative platform for experimenting with LLM inference optimizations, which can be run on a local desktop computer (Source and further information: GitHub[24]).

AI Assisted Knowledge Management (Mind Mapping)

AI-based information processing is an option but does not always lead to efficiency gains. This has already been pointed out in the context of hybrid search and is similarly true for some forms of knowledge management, such as the so-called Zettelkasten method, which is based on digital tagging. The Zettelkasten is a method of making short, atomic notes (containing the smallest unit of thought) that are linked together to form a system (Scheper, 2022). The associated information management technologies—such as tagging, referencing, backreference generation, knowledge graphing and mapping—support abductive logic, which involves inference based on assumptions. The software that supports this is known as Second Brain and has become popular among researchers (e.g. Obsidian, Logseq), especially those with AI copilot functionality. In addition, there are Second Brain AI agents that combine AI knowledge with RAG functions augmented with local documents to present direct results of semantic search (Abramdemski, 2019).

Reor is an AI-based desktop note-taking application: it automatically links logically related notes, answers questions about notes (with RAG), provides semantic search (on vector databases), and can generate AI flashcards. All elements are stored locally, and notes can be edited using an Obsidian-like markdown editor. (Source and further information: GitHub[25])

Drawing on chaos theory and graph theory, the Neurite explores the hidden patterns and complex relationships that shape creative thinking. The project combines fractals, mind mapping and AI to research, generate code, write, learn and visualise thinking. Its mission is to combine Zettelkasten and Mind Mapping. (Source and further information: GitHub[26])

Archyve includes an experimental Knowledge Graph feature, known as "Graph RAG". This functionality allows for Retrieval-Augmented Generation (RAG) by offering an API that queries the user's documents for relevant context. The Archyve client sends the user's prompt to Archyve, which then returns pertinent text snippets that the client can incorporate into its prompt for an LLM server (Source and further information: GitHub[27]).

---

[23] https://github.com/transformerlab/transformerlab-app

[24] https://github.com/kvcache-ai/ktransformers

[25] https://github.com/reorproject/reor

[26] https://github.com/satellitecomponent/Neurite

[27] https://github.com/nickthecook/archyve

Local AI for Mobile Devices

Personal LLM agents are a special type of LLM-based agents that deal with personal data and services running on local devices. They can be deployed on smart devices with limited resources and work with specially optimised, small AI models. The main task of personal LLM agents is to collect information from end users and process it on the local device. The necessary hardware requirements include a phone with at least 6 GB of RAM, several gigabytes of free back-up capacity, and a high-end chipset (on a chip) released after 2020, ideally a model manufactured after 2022.

Maid is an Artificial Intelligence distribution for mobile. Maid is a cross-platform (Desktop and Android), free and open-source application for local interfacing with llama.cpp models and remote interfacing with Ollama, Mistral, Gemini and OpenAI models (Source and further information: GitHub[28]).

ChatterUI is a mobile frontend for managing chat files and character cards. It supports different backends (e.g. text-generation-webui), connects to the LLM ggml via llama.rn and is able to run LLaMA models locally (Source and further information: GitHub[29]).

MLC LLM is a machine learning compiler and high-performance deployment engine for LLMs. The Android application (APK) is designed for devices with Snapdragon 8 Gen 2 chip but runs on other devices as well. The project's mission is to enable everyone to develop, optimize and deploy AI models natively on all platforms: iOS and Android (Source and further information: GitHub[30]).

Confichat is a lightweight, standalone chat interface designed for multiple platforms, emphasizing privacy and user autonomy. It integrates seamlessly with offline providers such as Ollama and LlamaCpp, ensuring that all interactions remain confined to your local machine or network, free from reliance on cloud services (Source and further information: GitHub[31]).

Rtranslator is an open-source real-time translation app for Android that runs locally. In walkie talkie mode, the smartphone microphone listens in two languages at once. This mode translates conversations between two people, parties. The app detects which language the other person is speaking, translates the voice into the other language, converts the text into audio, and then plays it back from the phone's speaker (Source and further information: GitHub[32]).

Sherpa is the deployment framework of the Next-gen Kaldi project, which can transcribe and generate speech using pre-trained models. k2-fsa/sherpa-onnx subproject enables speech-to-text, text-to-speech, voice recognition and voice activity detection (VAD) using last generation Kaldi and onnxruntime technologies.

---

[28] https://github.com/Mobile-Artificial-Intelligence/maid

[29] https://github.com/Vali-98/ChatterUI

[30] https://github.com/mlc-ai/mlc-llm

[31] https://github.com/1runeberg/confichat

[32] https://github.com/niedev/RTranslator

The FUTO Voice Input is an app that enables speech-to-text transcription on Android, integrating with third-party inputs (from keyboards or apps) using common speech-to-text APIs (Source: https://gitlab.futo.org/alex/voiceinput).

The following LLMs that work well on resource-constrained devices: llama3.2, phi3.5, gemma2:2b, qwen2.5:3b, nomic-embed-text, moondream (multimodal).

Artificial Intelligence Platforms
When listing Artificial Intelligence platforms, it is worth distinguishing between classic AI systems and Generative AI based on large language models, which are at the heart of today's technological revolution. Both classic and generative Artificial Intelligence solutions can significantly enhance various aspects of academic research, from data analysis to content generation, ultimately boosting productivity and effectiveness. Table 3.2 lists the most relevant platforms in these two categories at the time of writing.

4 Using AI Software from a Research Perspective (Application, LLM and External Data Source)

Second Brain AI Assistant
In everyday life, personal AI assistants have long been part of the portfolio of large IT service providers. Recently, attempts have been made to develop add-ons that monitor not only the user's daily computer or target device usage, but also their activities outside of it. While these features are a convenience, they raise a number of privacy issues. Conversely, the use of self-hosted monitoring solutions provides the opportunity to conduct activities in compliance with the relevant data protection principles, thus ensuring an acceptable privacy framework (Ifargan et al., 2024).

SurfSense is a tool that helps users to remember and organise the content they save while browsing the Internet. It acts as a personal knowledge graph (GraphRAG), allowing you to save and search web browser history and interact with natural language queries. Its main features are highly customised for your own operational needs, open-source and easy to install locally (Source and further information: GitHub[33]).

Adeus is an open-source, wearable AI tool that records conversations and voices real-time, writes down them and stores them on its own server. Users can then chat with Adeus using the app, which takes all relevant context into account for the best performance (Source and further information: GitHub[34]).

Screenpipe 24/7 is a nonstop screen and sound recording app, which creates a software development library to archive what the user has seen, said or heard. Both semantic and keyword-based searches are possible. Explore for information that the

[33] https://github.com/MODSetter/SurfSense

[34] https://github.com/adamcohenhillel/ADeus

Table 3.2 AI platforms (classic and generative ones) with features and advantages

Platform	Key features	Advantages
Classic AI		
Microsoft Azure	Cloud-based AI platform with AutoML, NLP, and computer vision	Scalable for large projects, integrates with Microsoft services, ideal for predictive modelling
TensorFlow	Open-source ML platform with deep learning support, model sharing, and visualization tools	Flexible, strong community support, extensive library of models, suited for custom AI model development
IBM Watson	Tools for data analytics, machine learning, and NLP	Excels in text analysis, automates workflows, integrates easily with enterprise tools
Amazon SageMaker	Managed ML platform covering full ML workflow from data prep to deployment	Simplifies ML lifecycle, cost-effective with Amazon cloud integration for large-scale projects
H2O.ai	AutoML-focused platform supporting various algorithms	Easy for non-coders, scalable, democratizes machine learning for researchers with diverse expertise
Generative AI		
ChatGPT	NLP for text generation, code assistance, and data analysis, with API integration	Useful for early research stages, summarizing text, and drafting, saving time on literature reviews
Grok	Unique access to posts on X, image and code generation	Up-to-date information, willingness to engage in controversial topics, playful and edgy tone
ScholarAI	Tools for citation, literature search, data extraction, and mapping connections	Automates citation management, streamlines literature reviews, suited for postgraduate researchers
Consensus	Summarizes academic papers, extracts insights from multiple studies	Reduces reading time, ideal for literature reviews and meta-analyses by synthesizing key findings
Gemini	Multi-modal AI for generating text and images based on natural language prompts	Assists in generating visual aids, valuable for interdisciplinary research with text and visuals
Perplexity AI	AI with human expertise for accurate, conversational summaries and insights	Provides concise answers, conversational interface for deeper topic exploration
DeepSeek	Open-source AI models with advanced reasoning capabilities, efficient resource utilization, and cost-effective deployment	Excels in mathematical reasoning, coding tasks, and complex problem-solving; accessible to a wide range of users due to open-source nature
Mistral AI	Open-source and commercial LLMs with API access via La Plateforme and chat-based AI assistant (Le Chat) with multilingual capabilities, up-to-date knowledge, web browsing, and multimodal capabilities	Open-weight models offer customization, support local deployment, and serve as a robust alternative to GPT models
SciSpace (former: Typeset.io)	End-to-end purpose-built workspace for academic research with AI-powered tools, including Chat with PDF, AI Writer, Paraphraser and Citation Generator	Speeds up literature review, enhances writing quality, automates citation management, and supports academic integrity

Source: authors' own compilation

user has forgotten or incorrectly categorized. It is free to use (Source and further information: GitHub[35]).

Retrieval-Augmented Generation and Hybrid Search

RAG is a method to improve the accuracy and reliability of Generative AI models with facts from external sources. RAG makes possible to store dense embeddings of documents in a vector database. The SPLADE (Source and further information: GitHub[36]) can also be used to generate sparse embeddings to enable hybrid search, which is the implementation of searching sparse and dense embeddings together, thus enabling more efficient information retrieval.

Verba is a fully customisable AI application that uses RAG to retrieve data. Different RAG frameworks, data types, chunking and retrieving techniques and LLM providers are available in the application according to the requirements of the user (Source and further information: GitHub[37]).

RAGFlow is an open-source RAG engine based on deep document understanding (Source and further information: GitHub[38]).

The pyLLMSearch—Advanced RAG's primary purpose is to retrieve local documents using LLM. For semantic search, it supports the "Retrieve and Re-rank" method, hybrid search and Hypothetical Document Embeddings (HyDE) capable search. It assists document collection and filtering of results by collection. If multiple queries are enabled, the original query is replaced by other three versions of the same query, allowing different views of the terminology to be cross-checked and selected (Source and further information: GitHub[39]).

Infinity Ais an AI-native database built for LLM applications that provides prompt hybrid searches between dense vectors, sparse vectors, tensors (multiple vectors), structured data and full texts. It provides intense help for various LLM applications including search, recommender, question answering, conversational AI, copilot, content generation and many other RAG applications (Source and further information: GitHub[40]).

DocsGPT is an open-source AI documentation assistant. DocsGPT simplifies finding information in different projects' documentation. By integrating powerful LLM models, developers can easily ask questions about their projects and get accurate answers (Source and further information: GitHub[41]).

The specialty of PrivateGPT is to show the user the results of the semantic search (i.e. search menu) (Source and further information: GitHub[42]).

[35] https://github.com/louis030195/screen-pipe

[36] https://github.com/naver/splade

[37] https://github.com/weaviate/Verba

[38] https://github.com/infiniflow/ragflow

[39] https://github.com/snexus/llm-search

[40] https://github.com/infiniflow/infinity

[41] https://github.com/arc53/DocsGPT

[42] https://github.com/zylon-ai/private-gpt

Local AI Web Search Engines
Generative AI is being increasingly integrated into large web search services. If we want to build a self-hosted web search engine, two main approaches can be followed: one is to build a locally-run search engine that covers a limited number of web pages (such as the YaCy project https://yacy.net/), while the other is to implement web crawler systems that produce Web ARChive files (such as the Scrapy framework, source and further information: GitHub[43]). A more up-to-date and resource-efficient alternative to these solutions are those that integrate the results of large search engines through the SearXNG metasearch engine, allowing AI applications running locally to operate more efficiently (Source and further information: GitHub[44]).

Farfalle is an open-source AI-driven search engine, which runs local LLMs or use cloud models. Searches can be performed with multiple search providers via the SearXNG metasearch engine (Source and further information: GitHub[45]).

Perplexica is also an open-source, Artificial Intelligence-driven search engine that uses advanced machine learning algorithms to refine results and provide definitive answers with referenced sources. Perplexica uses SearXNG as well, which is a meta-search engine to retrieve results, reorder them, and get the most relevant sources. Ollama can use local LLMs via a local API server (Source and further information: GitHub[46]).

The LLM_Web_search is a text generation WebUI extension that allows LLMs to search the web using duckduckgo. This action allows local LLMs to search the web by issuing a specific command. Finally, a combination of LangChain contextual compression and Okapi BM25 or SPLADE is used to extract the relevant parts of the web pages in the search results and appends the results to the model output (Source and further information: GitHub[47]) (Other information: https://python.langchain.com/docs/how_to/contextual_compression/).

Several major internet browsers, search platforms, such as Google, Bing, Perplexity and Brave, have previously integrated AI-based services into their services. However, these systems have been relatively quick to restrict the content of the services they offer, mainly due to hallucinations suggesting false information produced by AI. AI-based web search engines are able to search the web for relevant content and then summarise it to provide answers useful to the user. For these search results, the source of the web content is also referenced. The way web search engines work closely the technique, as RAG (Retrieval-Augmented Generation) based AI systems operate (Gao et al., 2023).

---

[43] https://github.com/scrapy/scrapy

[44] https://github.com/searxng/searxng

[45] https://github.com/rashadphz/farfalle

[46] https://github.com/ItzCrazyKns/Perplexica

[47] https://github.com/mamei16/LLM_Web_search

Literature Search Assistants

The performance of Generative Artificial Intelligence (GenAI) systems is closely linked to the quality and availability of the data they process. This is particularly important in contexts where AI makes use of literature resources. These systems often have access to open-source online bibliographic databases as well as to the full text of freely available scientific works. Such systems typically rely on the power of semantic search to find relevant literature and inform researchers in a conversational way (Corbin et al. (n.d.)).

OpenResearcher understands the questions researchers ask, searches the scientific literature, filters the information retrieved, and provides accurate and comprehensive answers. With access to the arXiv corpus, OpenResearcher can provide the latest scientific insights (Source and further information: GitHub[48]).

PaperQA2 is a high-precision, RAG (Retrieval-Augmented Generation) based system specialised in answering questions from scientific documents. It is able to contribute to scientific discourse by providing citations, summaries and identifying contradictions. One of the outstanding features of PaperQA2 is that it integrates and combines information from different metadata sources. Unpaywall can provide direct links to PDFs; Crossref delivers bibtex and Semantic Scholar keeps citation data (Sources and links: https://www.semanticscholar.org, https://www.crossref.org, https://unpaywall.org/). In this case, it is also possible to add documents manually.

Talk2Arxiv is an open-source RAG system specifically designed for PDFs of scientific articles. It employs GROBID (Source and further information: GitHub[49]) for efficient text extraction from PDF files. GROBID (GeneRation of BIbliographic Data) is an open-source software that generates bibliographic data and meta-information from scientific documents. Its main purpose is the automatic processing of scientific publications, especially in PDF format (Source and further information: GitHub[50]).

The ArXiv ChatGuru helps the discovery of scientific literature. The user submits a topic and ArXiv retrieves relevant articles from its collection of scientific papers (Source and further information: GitHub[51]).

The WikiChat is an improved RAG. It reduces the hallucination of large language models by querying data from the Wikipedia corpus. In addition to text, it supports querying structured data such as tables, infoboxes and lists (Source and further information: GitHub[52]).

The Semantic Scholar Semantic Reader is an advanced reading tool that can revolutionise scholarly reading by providing richer context. It uses AI to understand

[48] https://github.com/GAIR-NLP/OpenResearcher

[49] https://github.com/kermitt2/grobid

[50] https://github.com/evanhu1/talk2arxiv

[51] https://github.com/redis-developer/ArXivChatGuru

[52] https://github.com/stanford-oval/WikiChat

the structure of documents and uses Semantic Scholar's scholarly corpus as a data source (Sources:[53] https://www.semanticscholar.org/product/semantic-reader).

The STORM is one of the LLM systems that compiles quasi-Wikipedia articles based on an internet search. Co-STORM enables human collaboration with the LLM system by supporting more accurate information retrieval and knowledge systematisation. STORM cannot produce complex ready-to-use articles because the materials it produces often require further editing. Nevertheless, Wikipedia editors have found it useful as a first point of reference on a given topic (Source and further information: GitHub[54]).

Research Data Analysts: Data Analysis

AI applications are most often applied in different scientific fields to process such complex data sources that only specialised systems can handle efficiently. These data often contain large amounts of unstructured information, such as scientific articles, conference presentations, research reports and other publications.

Specialised AI applications are not only able to process raw data, but also to identify scientific context and relevant relationships. For example, Natural Language Processing (NLP) technologies enable systems to extract relevant information, identify keywords and understand the meaning behind scientific discourse.

The WARC-GPT is a test RAG pipe specifically designed to process web archive collections, i.e. WARC (Web ARChive) format files. WARC-GPT allows the creation of custom chatbots that use the content of web archives as a knowledge base. It offers a new approach to search, allowing full-text search coupled with a summary search function to discover information in web archives. WARC-GPT also supports visualisation of embeddings, helping users to better understand the complex structure of web archives and to use information more efficiently. With this solution, WARC-GPT contributes to easier access and processing of the knowledge contained in web archives (Source and further information: GitHub[55]).

Sparrow is an innovative open-source solution for efficient data extraction and processing (Optical Character Recognition <OCR> for document structure analysis) from various documents and images (Source and further information: GitHub[56]).

The FreeMoCap project is a free and open-source, hardware- and software-independent, low-cost research motion capture system. The application has unique GUI (Source and further information: GitHub[57]).

Domain-Specific LLMs

In the case of specific research, general LLMs do not provide adequate results, so it may be necessary to use domain-specific LLMs (Yan et al., 2024). About the StructLM: Large language models have difficulty interpreting and exploiting

[53] https://github.com/rayfok/scim

[54] https://github.com/stanford-oval/storm

[55] https://github.com/harvard-lil/warc-gpt

[56] https://github.com/katanaml/sparrow

[57] https://github.com/freemocap/freemocap

structured data. StructLM models, trained on a comprehensive instruction-annotation data set, achieve significant improvements in structured knowledge-based tasks over existing models. A series of models based on the Code-LLaMA architecture, known as StructLM, have been trained, with parameters ranging from 7B to 34B (Source: https://tiger-ai-lab.github.io/StructLM).

ColBERT is a fast and accurate search model that enables scalable BERT-based searches of large text collections in tens of milliseconds. ColBERT, a new text ranking model that adapts deep LMs (BERT) for efficient retrieval (Source and further information: GitHub[58]).

Autonomous Agents for Entire Scientific Research Workflow

The applications described in this section could be the most advantageous one, but also the most disturbing software, since they would automatically perform the writing of a literature-based paper on the selected topic, without human intervention. Research agents powered by a large language model automatically generate problems, methods and experimental designs while iteratively refining them based on the scientific literature. Around 80–90% of peer-reviewed publications (topic papers) are summarised without major errors, but as the complexity of the objectives increases, "human oversight" becomes critical to ensure accuracy. Software developers also believe that this standard of simple subject documentation must be exceeded. It is the abductive (inference based on assumptions) process of human reasoning that provides the creativity that drives scientific innovation, and this is what is missing from these summaries (topic papers). If we settle for the standard of topic papers, then this AI software is a threat to researchers. But if it is just purely used as a tool and assistance, it can be stimulus for human innovative ideas.

Data-to-paper is an automation framework that systematically guides interacting AI agents through the entire process of scientific research. It guides interactive LLMs and rule-based agents along traditional scientific methods, from data discovery and literature search, through annotated data and idea collection, to data analysis and interpretation, and finally to the creation of transparent, traceable, human-verifiable scientific publications. The platform is able to operate fully autonomously, but is also accessible through the Copilot application, which allows human control. The users can supervise, monitor and control the research process, define research objectives or allow the AI to generate and test hypotheses autonomously (Source and further information: GitHub[59]).

GPT Researcher is an LLM-based autonomous agent that conducts comprehensive research on any topic online. The agent is able to produce detailed, factual and unbiased research reports, with a focus on relevant sources and outlines, as well as other customisation options. The research agent driven by a large language model automatically generates problems, methods, and experimental designs while iteratively refining them based on the scientific literature. The recently released Plan-and-Solve and RAG connect information through a scientific graph. Local

[58] https://github.com/stanford-futuredata/ColBERT

[59] https://github.com/Technion-Kishony-lab/data-to-paper

documents and files can be searched and conduct hybrid research can be run using local and web-based sources (Source and further information: GitHub[60]).

The AI Scientist means a comprehensive system of AI agents that can conduct scientific research and discover new skills (Bostrom 2017). The AI Scientist supports fully automated scientific discovery and enables LLM base models to conduct research autonomously (Source and further information: GitHub[61]).

5 A Case Study: Elements Integrating into Standard Research Workflows

The previous sub-chapters have provided a selection of AI software from a specific research perspective. In the following, we aim to offer an example of a software and data source system that meets the principles formulated for the needs of research (Liu - Bates 2025). Several elements of the above process are applied as a test (pilot project) to the operation of the University of Miskolc's Library and Information Science AI (LIS-AI). The collection of relevant literature from different sources, not only from commercial databases, is a legal obligation and task of the university library. This will result in the creation of a scientific data repository of library AI, which will be managed using bibliographic reference management software.

AI software integrated into the scientific research process can greatly assist professionals in the analysis of research data (measurements, other structured data, etc.) (AI in Data Analysis). The integration of research data used by researchers into the literature allows AI to provide interpretations (Interpreting Research Outcomes with AI). The literature of the scientific data repository complements researchers' knowledge of the meaning of individual data, thanks to the analytical capabilities of AI (Table 3.3).

6 Conclusion

The use of AI in academic research has enormous potential. It is important to see it not just as a technical achievement, but as a well-designed and prepared system, running on locally hosted, open-source software and cost-effective hardware components. The system operators are not afraid that humans will be crowded out of research, but that the information generated by algorithms could outperform human performance in the age of AI (Jurafsky & Martin, 2022).

Generative AI is data-driven, while research is knowledge-intensive. Software supporting both activities is based on data, especially qualitative data. Building and

[60] https://github.com/assafelovic/gpt-researcher

[61] https://github.com/SakanaAI/AI-Scientist

Table 3.3 Workflow of AI tools for research

Phases	Desktop	Auton.Agents	On Premise	CloudProvider
1. Data collection	Jabref/Zotero + scite.ai API	GPT-researcher	Wikindx	Scite.ai 'smartcite'
2. AI: NLP	GPT4all, Alpaca, Speech Note	GPT-researcher	OpenWebUI	SciSpace Typeset.io
3. Search Web	Firefox Translator 132 + Page Assist YaCy	GPT-researcher	OpenWebUI	Perplexity.ai
4. Search LocalDocs	Recoll, GPT4all	GPT-researcher	OpenWebUI	SciSpace
5. Knowledge Man	Logseq +AIollama, +AssistSeq	GPT-researcher	Humhub	Obsidian tai_plugin
6. Writing	Lyx-semantic	GPT-researcher	Manubot AIEditor	SciSpace Typeset.io

1. Ollama API server: model Mixtral 8×7B (Mixture of Experts)
2. Hardware: high-performance workstation with: 2×20GB GPU NVLink, 128GB RAM, AVX512 CPU
Source: authors' compilation

maintaining institutional data assets is essential, including expert assembly of scientific data assets, which can be done by exploiting the legal framework provided by libraries. The creation of domain-specific scientific data repositories is fundamental. It is also true for Generative AI that the performance of the system is closely linked to the quality of the data integrated into it. As AI learns from the knowledge and data assets, any incorrect or inconsistent data can lead to erroneous conclusions (Liu - Bates 2025). In addition, a self-run research AI system requires users to constantly experiment, pay attention, provide feedback and evaluate.

Much of the history of AI has been built around deductive and inductive approaches. However, the basis of human intelligence is abduction, which is reasoning based on assumptions, and it is the source of creative and innovative ideas. Abductive reasoning can be defined as the cognitive ability to generate intuitions and hypotheses. This aspect makes the role of the human being in research indispensable (Maslej et al., 2024).

The integration of Artificial Intelligence into various disciplines is not intended to replace human efforts, but rather to enhance them, enabling more efficient workflows, deeper insights and the exploration of previously unimaginable solutions.

References

abramdemski. (2019). *The Zettelkasten method.* https://www.lesswrong.com/posts/NfdHG6oHBJ8Qxc26s/the-zettelkasten-method-1

Béchard, P., & Ayala, O. M. (2024). *Reducing hallucination in structured outputs via retrieval-augmented generation.* ArXiv E-Prints. https://doi.org/10.48550/arXiv.2404.08189

Biderman, D., Ortiz, J. G., Portes, J., Paul, M., Greengard, P., Jennings, C., King, D., Havens, S., Chiley, V., Frankle, J., Blakeney, C., & Cunningham, J. P. (2024). *LoRA learns less and forgets less*. ArXiv E-Prints. https://arxiv.org/abs/2405.09673v1

Bostrom, N. (2017). Superintelligence: Paths, dangers, strategies (Reprinted with corrections 2017). Oxford University Press.

Buolamwini, J. (2023). Unmasking AI: A story of hope and justice in a world of machines (First edition). Random House.

Corbin, T., Dawson, P., Nicola-Richmond, K., & Partridge, H. (n.d.). 'Where's the line? It's an absurd line': Towards a framework for acceptable uses of AI in assessment. Assessment & Evaluation in Higher Education, 1–13. https://doi.org/10.1080/02602938.2025.2456207.

Edge, D., Trinh, H., Cheng, N., Bradley, J., Chao, A., Mody, A., Truitt, S., & Larson, J. (2024). *From local to Global: A graph RAG approach to query-focused summarization*. ArXiv E-Prints. https://doi.org/10.48550/arXiv.2404.16130

Gao, L., Biderman, S., Black, S., Golding, L., Hoppe, T., Foster, C., Phang, J., He, H., Thite, A., Nabeshima, N., Presser, S., & Leahy, C. (2020). *The pile: An 800GB dataset of diverse text for language modeling*. ArXiv E-Prints. https://doi.org/10.48550/arXiv.2101.00027

Gao, Y., Xiong, Y., Gao, X., Jia, K., Pan, J., Bi, Y., Dai, Y., Sun, J., Wang, M., & Wang, H. (2023). *Retrieval-augmented generation for large language Models: A Survey*. ArXiv E-Prints. https://doi.org/10.48550/arXiv.2312.10997

Ifargan, T., Hafner, L., Kern, M., Alcalay, O., & Kishony, R. (2024). *Autonomous LLM-driven research from data to human-verifiable research papers*. ArXiv E-Prints. https://doi.org/10.48550/arXiv.2404.17605

Jiang, A. Q., Sablayrolles, A., Roux, A., Mensch, A., Savary, B., Bamford, C., Chaplot, D. S., de las Casas, D., Hanna, E. B., Bressand, F., Lengyel, G., Bour, G., Lample, G., Lavaud, L. R., Saulnier, L., Lachaux, M.-A., Stock, P., Subramanian, S., Yang, S., & Sayed, W. E. (2024). *Mixtral of experts*. ArXiv E-Prints. https://doi.org/10.48550/arXiv.2401.04088

Jurafsky, D., & Martin, J. H. (2022). *Speech and language processing: An introduction to natural language processing, computational linguistics, and speech recognition* (2nd ed., fifteenth impression, 2022). Pearson.

Li, J., Wang, S., Zhang, M., Li, W., Lai, Y., Kang, X., Ma, W., & Liu, Y. (2024). *Agent hospital: A simulacrum of hospital with evolvable medical agents*. ArXiv E-Prints. https://doi.org/10.48550/arXiv.2405.02957

Liu, D., & Bates, S. (2025). Generative AI in higher education: Current practices and ways forward. PDF

Maslej, N., Fattorini, L., Perrault, R., Parli, V., Reuel, A., Brynjolfsson, E., Etchemendy, J., Ligett, K., Lyons, T., Manyika, J., Niebles, J. C., Shoham, Y., Wald, R., & Clark, J. (2024). *Artificial intelligence index report 2024*. ArXiv E-Prints. https://doi.org/10.48550/arXiv.2405.19522

Russell, S., & Norvig, P. (2021). *Artificial intelligence, global edition. A modern approach* (4th ed.). Pearson Education. Retrieved from https://elibrary.pearson.de/book/99.150005/9781292401171

Scheper, S. P. (2022). *Antinet Zettelkasten: A knowledge system that will turn you into a prolific reader, researcher and writer*. Independently published.

Wang, J., Wang, J., Athiwaratkun, B., Zhang, C., & Zou, J. (2024). *Mixture-of-agents enhances large language model capabilities*. ArXiv E-Prints. https://doi.org/10.48550/arXiv.2406.04692

Wei, J., Wang, X., Schuurmans, D., Bosma, M., Ichter, B., Xia, F., Chi, E., Le, Q., & Zhou, D. (2022). *Chain-of-thought prompting elicits reasoning in large language models*. ArXiv E-Prints. https://doi.org/10.48550/arXiv.2201.11903

Yan, E., Bischof, B., Frye, C., Husain, H., Liu, J., & Shankar, S. (2024). Applied LLMs—What we've learned from A year of building with LLMs. *Applied LLMs*. https://applied-llms.org

Zhuang, A., Zhang, G., Zheng, T., Du, X., Wang, J., Ren, W., Huang, S. W., Fu, J., Yue, X., & Chen, W. (2024). *StructLM: Towards building generalist models for structured knowledge grounding*. ArXiv E-Prints. https://doi.org/10.48550/arXiv.2402.16671.

Part II
Practical Applications and Excellence in AI-Driven Research

Chapter 4
Using AI for Literature Review

Zhenyu Sun and Ruixue Yuan

Abstract With the rapid advances in generative AI, especially large language models (LLMs), the methodologies used in literature reviews have undergone significant transformation, enhancing both efficiency and quality. The chapter discusses how generative AI, particularly LLMs, strengthens the literature review process by refining key steps such as the formulation of research questions, the selection of relevant databases, and the development of effective search strategies. It highlights how LLMs support the extraction and synthesis of information, ensuring thorough analysis of diverse sources. Furthermore, the chapter considers the role of LLMs in organising and standardising data, leading to a more coherent and comprehensive synthesis of the literature. Concluding with an evaluation of current applications and future potential, the chapter presents practical case studies and step-by-step examples, offering students and researchers actionable insights on effectively leveraging AI to improve the depth, breadth, and quality of their literature reviews.

1 Introduction

1.1 The Essence of Literature Reviews

A literature review forms the foundation of academic scholarship, providing not merely a summary but a critical analysis and synthesis of existing research on a given topic. As a core element of scholarly work, literature reviews appear in various forms, from stand-alone papers to essential sections within theses, dissertations, and research articles. The primary function of a literature review is to present a

Z. Sun
University of Chinese Academy of Sciences, Beijing, China

R. Yuan (✉)
Beijing University of Chemical Technology, Beijing, China
e-mail: yuanruixue@buct.edu.cn

X. Zhou, H. Al-Samarraie (eds.), *Institutional guide to using AI for research*,
Advances in Artificial Intelligence in Education 2,
https://doi.org/10.1007/978-3-031-94809-1_4

comprehensive overview of the current state of knowledge, enabling both researchers and readers to understand the context, evolution, and key debates in the subject area.

The literature review process extends beyond description to include critical evaluation of existing research. This ensures that the research question is well-defined, the theoretical framework is sound, and the methodology is well-suited. Positioning the current study within the broader body of relevant literature, the literature review not only provides necessary context but also establishes the foundation for further research, typically preceding the methodology and results sections in scholarly works (Hart, 1998; Ridley, 2012; Machi & McEvoy, 2016).

In higher education, particularly at postgraduate levels, constructing a literature review is a crucial part of academic development. It plays a key role in the preparation of theses, dissertations, and journal articles, and is often a required component of research proposals. This step often serves as a preliminary approval for students embarking on dissertation or thesis work (Boote & Beile, 2005).

Literature reviews may also take the form of review articles—scholarly papers that consolidate existing knowledge, including substantive findings, theoretical contributions, and methodological advancements on a specific topic. These reviews, as secondary sources, do not introduce new experimental work but instead synthesise and critique existing research. Predominantly published in academic journals, they differ from book reviews in scope and depth and are an essential element across various academic disciplines.

The process of conducting a literature review is dynamic and continuous, contributing to various stages of empirical research. It encompasses a range of cognitive activities, as described by Bloom's revised taxonomy of the cognitive domain, including remembering, understanding, applying, analysing, evaluating, and creating (Anderson et al., 2000). This framework underscores the intellectual complexity involved in a comprehensive literature review, highlighting its significance in advancing academic discourse.

1.2 Types of Literature Reviews

Following the discussion on the critical role of literature reviews in academic scholarship, it is important to understand the various types of literature reviews, each with its own methodological and philosophical foundations. The concept of the systematic review, which gained prominence in the 1970s, has led to a distinction between two primary forms: narrative reviews and systematic reviews. Each type offers distinct approaches and serves different purposes within research synthesis (Grant & Booth, 2009).

Traditionally, the term "literature review" without further specification often refers to a narrative review. Narrative reviews are broad, interpretive syntheses of existing research and can be evaluative, exploratory, or instrumental. They provide a more personalised and subjective exploration of the literature, typically shaped by the author's expertise and the thematic focus of the field. This approach allows greater flexibility, as the selection of studies is more intuitive and aligned with the author's perspective. However, narrative reviews do not strictly follow predefined inclusion and exclusion criteria, nor do they engage in systematic assessments of study quality. Instead, they rely on qualitative analysis, offering an interpretive synthesis of findings within the broader context of the subject matter. While this method enables deeper thematic exploration, it is more susceptible to the author's subjective insights and potential biases (Baumeister & Leary, 1997).

In contrast, systematic reviews are designed to address specific research questions through a focused and rigorous approach. They aim to gather all relevant research evidence by employing a comprehensive and transparent process. The literature search in a systematic review is meticulously planned, utilising a predefined strategy that spans multiple databases and sources. This ensures inclusivity and reduces bias. Criteria for study inclusion and exclusion are explicitly defined from the outset, facilitating objectivity and reproducibility in the review process (Higgins & Green, 2011).

Systematic reviews place strong emphasis on assessing study quality, often employing standardised tools to evaluate the risk of bias in the research. This rigorous process ensures that only the most scientifically robust studies are included in the final synthesis. Moreover, systematic reviews often incorporate meta-analysis, a statistical method that combines data from selected studies, yielding more reliable and generalisable conclusions (Glass, 1976). Reporting for systematic reviews is guided by frameworks such as the PRISMA statement, which ensures transparency and completeness in presenting findings. Given the evolving nature of research, systematic reviews are updated periodically to incorporate new evidence (Moher et al., 2009).

1.3 Integration of Review Methodologies

Building on the distinction between narrative and systematic reviews, it is important to recognise that these methodologies, while different, are not mutually exclusive. Each offers unique strengths that can be leveraged depending on the research objectives. Narrative reviews allow for an individualised and flexible exploration of literature, shaped by the author's expertise and thematic focus. In contrast, systematic reviews aim to reduce subjectivity and bias through rigorous processes of study selection and evaluation, resulting in an objective synthesis that often relies on quantitative analysis (Higgins & Green, 2011).

Literature reviews can also be categorised as either "selective" or "rigorous." Narrative reviews, which are commonly found in theses and dissertations, tend to be more selective, reflecting the author's particular perspective and expertise. Conversely, systematic reviews, especially those published in peer-reviewed journals, adhere to strict standards such as the PRISMA guidelines. These guidelines require comprehensive searches across multiple databases and the use of predetermined protocols to ensure thoroughness, objectivity, and reproducibility (Moher et al., 2009).

In addition to these traditional approaches, integrative literature reviews have become a powerful method for generating new insights. This type of review not only synthesises existing literature but also critiques and builds upon it to offer new perspectives on a given topic. Scholars have identified several distinct types of literature reviews, including scoping reviews, systematic or integrative reviews, meta-narrative reviews, critical reviews, meta-analyses, and mixed research syntheses. Each serves a specific purpose and is chosen based on factors such as the review's objectives and the nature of the subject matter (Torraco, 2005).

By integrating these various methodologies, researchers can better align their review strategies with their research questions. This approach enriches academic discourse by offering both depth and breadth in the analysis of existing research, ensuring that the literature review serves as a comprehensive foundation for future investigations (Boote & Beile, 2005).

1.4 The Role of AI in Enhancing Literature Review

In the evolution of academic research, artificial intelligence (AI) is increasingly transforming the literature review process. Platforms such as ChatGPT are reshaping how researchers identify, analyse, and synthesise existing literature. By incorporating AI into these processes, researchers can explore their fields more thoroughly, uncovering methodologies, techniques, limitations, and knowledge gaps that might otherwise be overlooked. This enhanced capability facilitates a deeper understanding of current studies while simultaneously encouraging new research directions, advancing knowledge across various domains.

This chapter discusses how AI-assisted tools like ChatGPT improve the efficiency, scope, and depth of literature reviews. By automating tasks such as data extraction and synthesis, AI allows researchers to manage large volumes of information more efficiently. Additionally, AI can assist in cross-referencing studies, identifying relevant citations, and suggesting connections between disparate research areas. This broadens the researcher's intellectual scope and helps form a more comprehensive view of the subject matter.

However, the integration of AI into academic research comes with its challenges. Ethical and methodological considerations need to be addressed carefully to ensure that AI tools are applied in ways that uphold the principles of research integrity and scholarly communication. This chapter will also consider these challenges, offering practical guidelines for incorporating AI into literature reviews while maintaining high standards of academic rigour and ethical responsibility (Floridi & Cowls, 2019). By the end of this chapter, our future researchers will gain insights into how AI can enhance the quality and impact of their literature reviews, positioning them to lead academic innovation.

1.4.1 Steps of AI-Assisted Literature Review

This chapter provides a detailed introduction to conducting a literature review with the aid of AI, using the philosophy of science as a case study. Science is not just a repository of empirical facts and theories; it is deeply influenced by philosophical inquiry. By examining the philosophical foundations of scientific practice, this review aims to illuminate the methodological, epistemological, and ontological aspects that shape our understanding of scientific knowledge. For this case study, we will consider the following research question in the subject of Philosophy of Science:

Topic**:** What is the essence of gauge structure in quantum field theory?

This section outlines a structured approach to utilising AI technologies for compiling and analysing literature within the philosophy of science. The discussion will explore how AI can streamline various stages of the review process, including the formulation of research questions, selection of bibliographic databases, optimisation of search algorithms, screening of literature, and extraction and synthesis of data. The result is a comprehensive review that reflects the current state of discourse while paving the way for future philosophical investigations.

As we will see, the process of constructing an AI-assisted literature review is not merely mechanical. It is an intellectual endeavour that enhances philosophical dialogue and deepens our engagement with complex scientific concepts. This journey demonstrates the potential of AI to elevate philosophical inquiry in science.

The process of writing literature reviews with the help of generative AI (GenAI) typically involves six key steps, each of which contributes to constructing a thorough and coherent review. These steps are not isolated actions but interrelated phases that guide researchers from clarifying their inquiry to synthesising complex philosophical debates. This structured approach ensures that the review is both comprehensive and insightful, providing a solid foundation for further academic exploration.

The first and foundational step is *clarifying the research questions*. This stage is essential as it establishes the focus and scope of the literature review. Well-defined questions help guide the selection of relevant literature, ensuring that the review remains coherent and aligned with the overarching themes of the research. For instance, in the context of quantum field theory, research questions might focus on the theoretical frameworks and interpretations of gauge symmetries, shaping the entire inquiry. By framing these questions clearly, the researcher ensures that the literature review will address the core philosophical debates pertinent to the subject.

Once the research questions are in place, the next step is *selecting appropriate bibliographic databases*. This involves identifying the most relevant sources of literature that align with the research's thematic focus. For philosophical research, databases such as PhilPapers, JSTOR, and Google Scholar are invaluable as they provide access to a broad range of academic work. Careful selection of these databases ensures the inclusion of high-quality sources, laying the groundwork for a rigorous and relevant literature review.

The third step, *developing a search algorithm*, is critical for efficiently identifying literature. A well-crafted algorithm includes a combination of keywords and subject terms that reflect the key themes of the research. In philosophy of science, for example, terms like "gauge theory," "realism vs. anti-realism," and "induction problem" are essential for capturing a focused yet comprehensive range of literature. By designing this algorithm carefully, researchers can ensure that their search yields a balanced selection of sources relevant to their inquiry.

After compiling the initial body of literature, the next step is *literature screening and quality assessment*. This involves critically evaluating each source for its relevance and scholarly rigor. Systematic criteria are applied to assess the philosophical arguments, methodologies, and contributions to the existing discourse. By rigorously filtering sources, the researcher ensures that only the most pertinent and reliable materials are included in the review, enhancing the quality and focus of the analysis.

Following the screening, the process moves on to *data extraction and standardisation*. Key information from literature—such as author names, publication details, central arguments, and methodologies—needs to be organised in a consistent format. This standardisation allows for easier comparison and synthesis across diverse perspectives, ensuring that the review remains coherent while integrating various philosophical positions.

The final step involves *synthesis and meta-analysis*. At this stage, the researcher synthesises the extracted data, identifying areas of convergence and divergence in the philosophical debates. While meta-analysis is less common in philosophical work due to its qualitative nature, thematic synthesis can reveal broader trends and ongoing debates.

By following this systematic approach, researchers can effectively leverage AI tools to create a literature review that not only addresses current debates but also opens pathways for future inquiry. The subsequent sections will offer a detailed exploration of each step, providing practical guidance on how to implement these strategies in AI-assisted literature reviews.

2 Research Question Clarification

2.1 The Importance of Domain Knowledge

In the pursuit of scholarly knowledge, each discipline establishes core frameworks that underpin the analysis and resolution of problems specific to that field. These frameworks are vital as they provide structure, guide research efforts, and ensure that inquiries are systematic and rigorous. When employing AI to assist with a literature review, the adoption of such frameworks becomes even more crucial. AI algorithms require a strong grasp of domain-specific knowledge to effectively identify, categorise, and synthesise relevant information. Thus, domain expertise is indispensable for producing reliable and meaningful results.

For example, in the field of medicine, the PICOS framework (Population, Intervention, Comparison, Outcome, Study design) is commonly used to structure research questions and guide the search for evidence-based practices. This systematic approach ensures that the research process remains focused and yields relevant outcomes. Studies have shown that using structured frameworks, such as PICOS, improves the precision of literature searches and enhances the relevance of the retrieved studies (Liberati et al., 2009).

Similarly, in economics, empirical frameworks are used to analyse data and draw conclusions about economic phenomena. These frameworks typically involve assumptions, theoretical models, and statistical methods that guide hypothesis testing and the evaluation of economic theories. Such a structure is crucial for AI tools to identify relevant studies and synthesise data effectively. The use of empirical models enables researchers to streamline the process, ensuring that AI captures essential patterns and insights from the literature (Angrist & Pischke, 2009).

When addressing complex questions in the philosophy of science, such as "What is the essence of gauge structure in quantum field theory?", a structured approach is equally important. In this context, a philosophical framework might include several key elements:

1. **Conceptual Analysis**: Begin by clearly defining key concepts, such as "gauge" and "quantum field theory," to ensure a shared understanding of the terminology used in the discourse.
2. **Historical Context**: Explore the historical development of the concept within physics, tracing its evolution and the philosophical debates it has sparked over time.
3. **Interdisciplinary Integration**: Acknowledge the interdisciplinary nature of the question by integrating insights from physics, mathematics, and philosophy to form a comprehensive analysis.
4. **Philosophical Frameworks**: Identify and examine various philosophical frameworks that could be applied to the question, such as realism, empiricism, and the nature of scientific explanation.
5. **Debate Identification**: Identify and explore existing debates within the philosophy of science that are relevant to gauge structures, such as the role of symmetry in physical laws or the ontological implications of gauge symmetry breaking.
6. **Critical Evaluation**: Critically assess the literature by evaluating the strengths and weaknesses of different philosophical arguments and their implications for understanding gauge structures.
7. **Synthesis and Reflection**: Synthesize the findings from the literature review, reflecting on how they contribute to the broader philosophical discourse on the nature of gauge structures in quantum field theory.

By applying this structured framework when using AI to assist with a literature review in the philosophy of science, researchers can ensure a systematic and comprehensive exploration of the topic. AI enhances the ability to delve into the complexities of the subject, facilitating deeper philosophical inquiry and broadening the scope of understanding.

2.2 Prompt Design and Dialogue Workflow

In the following sections, we will demonstrate how to effectively use LLMs to assist in completing each step of the literature review process outlined earlier. By carefully crafting precise prompts and following a structured dialogue workflow, researchers can enhance both the efficiency and depth of their inquiry through the application of LLMs.

2.2.1 Conceptual Analysis

The first step in conducting a thorough literature review is to ensure a clear and accurate understanding of the key concepts involved in the research question. This is essential for building a solid foundation for the review and for engaging meaningfully with the relevant literature.

Prompt 2.2.1 (Conceptual Analysis)
Please help me complete the writing of a literature review. My research topic is: "What is the essence of gauge structure in quantum field theory?"

Please assist me with the conceptual analysis of this topic to ensure that there is a shared understanding of the core terms within the discourse.

Here is the response provided by the LLM:

Conceptual analysis of the research question "What is the essence of gauge structure in quantum field theory?" involves the following key steps:

1. **Definition of Key Terms:** Begin by defining "gauge" within the context of physics. Gauge theories are field theories that remain invariant under continuous transformations, known as gauge transformations, which do not affect physical observables.
2. **Quantum Field Theory Context:** Quantum field theory (QFT) is the framework for constructing quantum models of subatomic particles and their interactions, forming the foundation of modern particle physics.

3. **Gauge Symmetry:** Gauge symmetry refers to a symmetry of the Lagrangian in a physical system under a group of transformations, which gives rise to conserved quantities and underpins fundamental forces.
4. **Physical Interpretation:** Gauge structures are more than mathematical constructs; they have significant physical interpretations, linking invariance under transformations to the identification of forces and particles.
5. **Philosophical Implications:** Gauge structures raise deep philosophical questions about the nature of physical reality, symmetry, and the relationship between mathematics and the physical world.
6. **Relevance to QFT:** Gauge structures are integral to the formulation of quantum field theories, particularly in describing the electromagnetic, weak, and strong forces through gauge symmetries.
7. **Clarity and Precision:** Ensuring conceptual clarity and precision is crucial for avoiding ambiguity in discourse, especially when engaging with the philosophical implications of gauge structures in QFT.

2.2.2 Historical Contextualization

After establishing a clear conceptual framework, the next step is to situate the concept within its historical context. Understanding the historical development of gauge structures in quantum field theory allows for a deeper appreciation of its significance and the philosophical debates it has inspired.

Prompt 2.2.2 (Historical Contextualization)
To enhance my research question, "What is the essence of gauge structure in quantum field theory?", please conduct a historical contextualisation by investigating the evolution of this concept in physics.

The LLM's response goes as follows:

Historical contextualisation for the research question "What is the essence of gauge structure in quantum field theory?" involves tracing the development of gauge symmetry and its integration into the broader framework of physics, as well as the philosophical discussions it has prompted:

1. **Early Developments:** The concept of gauge invariance was introduced by Hermann Weyl in the early twentieth century, initially as an attempt to unify gravity and electromagnetism. Although this early idea was later revised, it laid the groundwork for further developments.
2. **Quantum Mechanics and Electromagnetism:** Gauge invariance became fundamental in quantum mechanics, particularly through the development of quantum electrodynamics (QED), where gauge symmetry played a crucial role in describing the electromagnetic interaction.

3. **Yang-Mills Theory:** In the 1950s, Chen Ning Yang and Robert Mills extended gauge symmetry to non-abelian groups, leading to the formulation of non-abelian gauge theories, which became the basis for the strong and weak nuclear forces.
4. **Standard Model:** The unification of electromagnetic and weak forces was achieved in the 1960s and 1970s through the electroweak theory, contributing to the Standard Model of particle physics.
5. **Gauge Symmetry Breaking:** Spontaneous symmetry breaking, along with the Higgs mechanism, explains particle masses, adding depth to both physical and philosophical discussions on symmetry.
6. **Current Trends:** Current research continues to explore the implications of gauge theory in areas such as the unification of forces, quantum gravity, and cosmology.

By examining the historical development of gauge theory, along with its philosophical implications, a deeper understanding of gauge structures in quantum field theory is achieved, contributing to the broader discourse on the nature of physical reality.

2.2.3 Interdisciplinary Integration

After establishing a conceptual framework and understanding the historical context, the next step in constructing a literature review on a complex topic like "the essence of gauge structure in quantum field theory" is to incorporate insights from multiple disciplines. An interdisciplinary approach is essential for developing a comprehensive and nuanced analysis, as it draws on the strengths and perspectives of different fields. This not only deepens the understanding of the research question but also broadens the scope of the inquiry.

Prompt 2.2.3 (Interdisciplinary Integration)
In this step, we will perform an interdisciplinary integration by synthesising perspectives from physics, mathematics, and philosophy. This approach is necessary to achieve a well-rounded analysis that fully addresses the breadth of the topic.

Given my research question, "What is the essence of gauge structure in quantum field theory?", please proceed with the interdisciplinary integration.

LLM Response:

Interdisciplinary integration for the research question "What is the essence of gauge structure in quantum field theory?" requires synthesising insights from physics, mathematics, and philosophy to develop a holistic understanding:

1. **Physics Perspective:** Examine the physical implications of gauge structures, focusing on their role in formulating the fundamental forces described by the Standard Model and their relevance to particle interactions.

2. **Mathematical Perspective:** Investigate the mathematical foundations of gauge theories, including group theory, differential geometry, and topology, which are essential for describing gauge symmetries and the fields associated with them.
3. **Philosophical Perspective:** Explore the philosophical dimensions of gauge structures, addressing questions about the nature of physical laws, the interpretation of quantum mechanics, and the ontological status of gauge symmetries.
4. **Historical Context:** Consider the historical development of gauge theory, recognising the contributions from multiple disciplines that have shaped its evolution and current understanding of physical phenomena.
5. **Conceptual Synthesis:** Integrate concepts from physics, mathematics, and philosophy to create a cohesive narrative about gauge structures, illustrating how each field contributes to a more complete understanding of the topic.
6. **Methodological Considerations:** Acknowledge the distinct methodological approaches of each discipline, such as experimental verification in physics, formal proofs in mathematics, and conceptual analysis in philosophy, and explore how these methods complement each other in the study of gauge structures.
7. **Current Research and Debates:** Identify ongoing interdisciplinary research and debates surrounding gauge structures, including theoretical advancements, experimental challenges, and differing philosophical interpretations.
8. **Future Directions:** Reflect on how interdisciplinary integration can inform future research directions, potentially leading to new insights or unifying principles in the understanding of gauge structures.

By engaging in interdisciplinary integration, the research question is approached from multiple angles, fostering a richer and more nuanced exploration of the essence of gauge structure in quantum field theory. This holistic approach not only deepens the analysis but also opens up new avenues for research and discovery.

2.2.4 Philosophical Frameworks

After integrating interdisciplinary insights, the next natural step is to apply various philosophical frameworks to the research question. This approach allows for a deeper exploration of the nature and significance of gauge structures in quantum field theory, providing insights into the foundational questions and assumptions underlying scientific theories.

Prompt 2.2.4 (Philosophical Frameworks)
In this step, we will engage with different philosophical frameworks to examine the concept of gauge structures from multiple philosophical perspectives. This will contribute to a richer and more nuanced understanding of the research question.

Given my research question, "What is the essence of gauge structure in quantum field theory?", please proceed with an analysis based on relevant philosophical frameworks.

The LLM's response goes as follows:

Addressing the research question, "What is the essence of gauge structure in quantum field theory?" through philosophical frameworks involves considering the following perspectives:

1. **Realism:** Explore whether gauge symmetries represent objective features of the physical world or if they are merely useful constructs within theoretical descriptions.
2. **Empiricism:** Investigate the role of empirical evidence in shaping and validating gauge theories. Examine how observations and experiments influence our understanding of gauge structures and their physical significance.
3. **Scientific Explanation:** Consider how gauge theories contribute to scientific explanations of physical phenomena. In particular, gauge principles underlie the Standard Model, providing explanations for fundamental forces and particle interactions.
4. **Instrumentalism vs. Theoretical Pluralism:** Reflect on whether gauge theories should be viewed as predictive tools or as revealing different aspects of a pluralistic theoretical landscape, where multiple interpretations coexist.
5. **Underdetermination of Theories:** Discuss the possibility that empirical data may support multiple theoretical frameworks and how this relates to the selection of certain gauge theories over others.
6. **Philosophy of Mathematics:** Given the reliance of gauge theories on advanced mathematical structures, explore the relationship between mathematics and physical reality, focusing on how mathematical entities like groups and fields describe physical phenomena.
7. **Causality and Determinism:** Gauge theories often involve elements of nondeterminism, especially in the context of quantum interactions. Analyse the philosophical implications of indeterminacy and causality in relation to gauge structures.
8. **Metaphysics of Quantum Mechanics:** Gauge structures are intrinsic to quantum field theory, which introduces its own metaphysical questions. Examine how quantum mechanics influences our understanding of space, time, and matter.
9. **Interpretations of Quantum Field Theory:** Consider how different interpretations of quantum field theory—such as the Copenhagen interpretation, many-worlds interpretation, or pilot-wave theory—affect the philosophical understanding of gauge structures.

By engaging with these philosophical frameworks, we can deepen the investigation into the essence of gauge structures in quantum field theory. This analysis not only enhances our understanding of their scientific role but also reveals broader philosophical implications.

2.2.5 Debate Identification

After examining the philosophical frameworks, the next crucial step for literature review in the field of philosophy is to identify key debates within the philosophy of science that are relevant to gauge structures. Identifying these debates helps to position your research question within the broader discourse, highlighting areas of contention and divergent interpretations in the field.

Prompt 2.2.5 (Debate Identification)
In this step, we will engage in the identification of existing debates within the philosophy of science that intersect with the concept of gauge structures. This will allow us to position your research within ongoing discussions and critically engage with the literature.

Given my research question, "What is the essence of gauge structure in quantum field theory?", please proceed with the corresponding debate identification.

The LLM's response goes as follows:

Identifying key debates for the research question, "What is the essence of gauge structure in quantum field theory?", involves exploring several philosophical discussions that intersect with the concept of gauge structures:

1. **Nature of Gauge Symmetry:** Debate whether gauge symmetries are fundamental aspects of the universe or emergent properties arising from current theoretical frameworks.
2. **Ontological Status:** Analyse the ontological status of gauge bosons and gauge fields. Are they real entities within the fabric of the universe, or merely conceptual tools aiding our understanding of physical phenomena?
3. **Epistemological Implications:** Explore the epistemological consequences of gauge symmetry, particularly how it shapes our understanding of the limits of scientific knowledge and the standards for what counts as an adequate scientific explanation.
4. **Interpretations of Quantum Mechanics:** Consider how different interpretations of quantum mechanics, such as the Copenhagen interpretation, Many-Worlds interpretation, or De Broglie-Bohm theory, influence philosophical debates about gauge structures.
5. **Unification of Forces:** Delve into the debates surrounding the unification of fundamental forces, focusing on the role of gauge theories in this effort, and the philosophical challenges and implications of such unification attempts.
6. **Quantum Field Theory and Reality:** Reflect on debates regarding the relationship between quantum field theory and physical reality, particularly how gauge symmetry manifests in the physical world.

7. **Gauge Freedom and Determinism:** Engage with philosophical debates about gauge freedom and its implications for determinism and predictability in quantum field theory.
8. **Philosophy of Mathematics:** Explore how the role of mathematics in formulating gauge theories contributes to ongoing debates about the relationship between mathematical structures and physical reality.
9. **Scientific Realism and Anti-Realism:** Examine how the broader debate between scientific realism and anti-realism plays out in the context of gauge theories, especially in relation to whether theoretical entities like gauge bosons exist independently of the theories that describe them.

By identifying these debates, we can better grasp the philosophical landscape surrounding gauge structures in quantum field theory, allowing for a deeper and more critical engagement with the research question.

3 Database Selection and Search Algorithm Development

The pursuit of scholarly knowledge is often compared to navigating a vast ocean of literature, where the precision of your research question serves as the compass guiding your journey. The importance of the "Research Question Clarification" step cannot be overstated, as it not only defines the scope of your inquiry but also enhances the efficiency with which relevant literature can be retrieved from bibliographic databases. As we progress to the task of selecting appropriate databases, it is vital to consider both specialised repositories that focus on specific domains and comprehensive databases that cover a wider range of scholarly discourse.

In this section, we will explore the strategic selection of databases that are rich in philosophical literature, such as PhilPapers, JSTOR, and Google Scholar, alongside other indices that may provide crucial insights into our research question: "What is the essence of gauge structure in quantum field theory?" This selection process is a critical methodological step, laying the groundwork for a rigorous and insightful literature review. By identifying the most relevant databases, we aim to streamline the search for literature that elucidates the philosophical foundations of gauge structures in quantum field theory, ensuring that the inquiry is both comprehensive and deeply informed.

3.1 Bibliographic Database Selection

Selecting the right bibliographic databases is essential for conducting a comprehensive literature review. The clarity of your research question acts as a guiding principle, steering you through the extensive array of scholarly resources available. Building on previous discussions, we now turn to the strategic selection of databases that will serve as the foundation for exploring the essence of gauge structure in quantum field theory.

Prompt 3.1.1 (Database Comparison)
Following from our prior discussions, the next step for your research inquiry, "What is the essence of gauge structure in quantum field theory?" is to engage in "Bibliographic Database Selection."

Please enumerate a comprehensive list of philosophical literature databases and provide an analysis of their respective advantages and disadvantages.

The LLM's response goes as follows:

For the research question "What is the essence of gauge structure in quantum field theory?", selecting the appropriate databases is crucial to ensure a comprehensive literature review. Here are several key philosophical databases, along with their advantages and potential limitations:

1. **PhilPapers**
   - *Advantages:* PhilPapers is a comprehensive index of philosophy, including articles, books, and conference papers. It offers a user-friendly search interface and regular updates.
   - *Disadvantages:* While extensive, it may lack some historical texts or recent publications not yet indexed.
2. **JSTOR**
   - *Advantages:* JSTOR provides access to a vast collection of academic journals, books, and primary sources. It is particularly strong in the humanities and social sciences, including philosophy.
   - *Disadvantages:* Its coverage is retrospective, and some content may require a subscription or institutional access.
3. **Google Scholar**
   - *Advantages:* Google Scholar indexes a wide array of scholarly literature across many disciplines. It is free to use and often includes citations for conference papers, theses, and other materials not indexed elsewhere.
   - *Disadvantages:* Quality may vary, and it may include non-peer-reviewed documents. It also lacks advanced search features found in specialised databases.
4. **Project MUSE**
   - *Advantages:* MUSE provides full-text access to current content from renowned humanities and social sciences journals, including philosophy.

- *Disadvantages:* Its focus is more on the humanities, and it may not be as comprehensive for philosophy of science as other databases.

5. **ERIC (Educational Resources Information Center)**

   - *Advantages:* Although primarily an education database, ERIC contains philosophy-related literature, especially in educational theory and philosophy of education.
   - *Disadvantages:* It may not be as relevant for the specific themes of quantum field theory and gauge structures.

6. **Philosopher's Index**

   - *Advantages:* This index covers a wide array of philosophical literature and is updated regularly, offering extensive coverage of journals and books.
   - *Disadvantages:* Access is often restricted to subscribers, and it may not include as many primary sources as some other databases.

7. **Web of Science**

   - *Advantages:* Web of Science offers a comprehensive citation database across multiple disciplines, including philosophy. It allows for citation tracking and analysis of how ideas have developed over time.
   - *Disadvantages:* It can be expensive to access and may not be user-friendly for those unfamiliar with citation databases.

8. **SpringerLink**

   - *Advantages:* SpringerLink contains a large collection of philosophy books and journals, many of which are high-quality publications.
   - *Disadvantages:* Access is typically subscription-based, and the focus is more on published works, rather than grey literature.

9. **Oxford Scholarship Online**

   - *Advantages:* Offers peer-reviewed works from Oxford University Press, including a significant number of philosophy titles.
   - *Disadvantages:* Access is usually restricted to subscribers, and its scope is more limited compared to broader databases.

When selecting databases for your literature review, consider factors such as the scope of coverage, relevance to your specific research question, accessibility, and availability of full-text resources. Combining these databases will provide a multi-faceted view of the philosophical discourse surrounding gauge structures in quantum field theory.

The process of database selection is not merely a procedural task but a critical scholarly decision. Balancing AI-generated recommendations with personal investigation and expert guidance is essential. While AI can efficiently suggest databases based on algorithmic analyses, the researcher's discernment, informed by years of intellectual engagement and advice from senior scholars, plays a crucial role in selecting resources that are both relevant and authoritative.

Incorporating AI suggestions, personal research, and expert insight ensures a robust and well-rounded approach to database selection. This strategy not only enhances the efficiency of your literature review but also grounds it in both data-driven insights and deep scholarly understanding.

As you navigate the complex landscape of philosophical literature in your exploration of gauge structures in quantum field theory, a balanced approach that combines AI assistance with scholarly experience will result in a more comprehensive and insightful literature review. This method ensures that your research is both thorough and analytically rigorous, laying a strong foundation for further inquiry.

3.2 *Search Suggestion Heuristics*

After selecting the appropriate databases, the next crucial step is to develop and optimise a search strategy that yields relevant and comprehensive results. This subsection uses the PhilPapers database as a case study to demonstrate how to create and refine a literature search algorithm tailored to the research question, "What is the essence of gauge structure in quantum field theory?"

Prompt 4.2.1 (Search Suggestion Heuristics)
Building on our previous discussions on "Research Question Clarification" and "Bibliographic Database Selection," the next step is to formulate a strategic approach for conducting a literature search in the PhilPapers database relevant to our research topic.

The LLM's response goes as follows:

To create an effective search strategy in the PhilPapers database for the research question "What is the essence of gauge structure in quantum field theory?", the following steps should be considered:

1. **Keyword Specification**: Start by identifying precise keywords that capture the core concepts of your research. For this inquiry, terms such as "gauge theory," "quantum field theory," "philosophy of physics," and "scientific explanation" should be prioritised
2. **Advanced Search Techniques:** Use PhilPapers' advanced search functions, incorporating Boolean operators (AND, OR, NOT) to combine keywords. This helps refine the search results and ensures that the query aligns with the focus of your research.
3. **Category Filtering:** PhilPapers organises publications by subject area. Apply filters to narrow your search to categories like "Philosophy of Physics," "Quantum Mechanics," or "Metaphysics" to improve the relevance of your results.

4. **Author Expertise:** Identify leading scholars in the philosophy of science who have made significant contributions to the field. Targeting works by these prominent authors can provide a solid foundation for your literature review.
5. **Publication Date Range:** Since gauge theories have evolved significantly over time, define a date range that captures both the historical development and recent advancements in the field. This ensures your review is both comprehensive and up-to-date.
6. **Citation Analysis:** PhilPapers offers citation data. Use this feature to identify highly cited works, as these are likely seminal texts in discussions surrounding gauge structures in quantum field theory.
7. **Abstract Review:** Before retrieving full texts, review abstracts to quickly assess whether the articles are relevant to your research question. This step can save time and resources by filtering out less pertinent studies early on.
8. **Iterative Search Refinement:** The literature search is typically iterative. Based on your initial search results, refine your search terms and strategies to ensure that your queries better capture relevant literature.
9. **Record Management:** Maintain a record of your search history, including the keywords and search strings used, to ensure reproducibility and facilitate adjustments in subsequent searches.
10. **Engagement with the Community:** PhilPapers also serves as a platform for scholarly discussions. Engage with the community by following ongoing debates and discussions, which can provide additional insights and potential leads for your literature review.

3.3 Search Algorithm Development

After selecting the appropriate databases and identifying key strategies for conducting a literature search, the next step involves developing and executing a precise search algorithm. This process is essential for extracting relevant, high-quality literature from your chosen database, in this case, PhilPapers.

In this phase, we will perform a preliminary search based on the AI's recommendations. It is important to recognise that effectively guiding the AI's responses—referred to as "interpolation"—can significantly improve the relevance of its suggestions. By iteratively refining the search strategy and incorporating AI-generated insights, the search can be better aligned with the specific nuances of your research question. This approach not only optimises the retrieval of pertinent scholarly works but also ensures that the results are closely tailored to the objectives of the literature review.

Prompt 3.3.1 (Plain Search Algorithm)
Please provide search keywords (e.g., "gauge symmetry"), field filter terms (e.g., "philosophy of physics"), and key authors (e.g., James Weatherall) for our research question in the PhilPapers Database.

The LLM's response goes as follows:

For the research question "What is the essence of gauge structure in quantum field theory?", here are suggested keywords, field filters, and key authors for conducting a search in the PhilPapers database:

Keywords	Gauge Theory, Gauge Symmetry, Quantum Field Theory
	Scientific Explanation
Field Filters	Philosophy of Physics, Quantum Mechanics
	Metaphysics, Epistemology, Philosophy of Science

Key Authors:

- James Weatherall (as suggested)
- Tim Maudlin (known for his work on the interpretation of quantum mechanics)
- Michael D. G. Redhead (for contributions to the philosophy of quantum field theory)
- Christopher A. W. Hill (for work on the philosophy of physics and quantum field theory)
- Robert Oerter (for writings on the mathematical physics of gauge theory)

When conducting the search, it is advisable to use these keywords in combination with field filters to narrow the results to the most relevant literature. Additionally, focusing on works by key authors who have made significant contributions to the philosophy of physics and quantum field theory will provide a more authoritative and detailed perspective on the topic.

Reviewing abstracts and citation counts during the search process will also help to assess the relevance and impact of the articles, ensuring that the literature review is both comprehensive and insightful.

3.4 *Refining and Optimizing the Search Algorithm*

Developing an effective search algorithm is essential for conducting a thorough and precise literature review. An optimised search strategy ensures that the most relevant and high-quality scholarly works are identified, thereby enriching the depth and rigour of your research. This process requires careful planning, iterative refinement, and a solid understanding of both the research question and the functionality of the chosen database.

While AI and LLM tools like ChatGPT can assist in generating initial search strategies, it is crucial to critically assess and refine these suggestions to ensure they meet the specific needs of your inquiry. Collaborating with experienced information scientists or librarians can further enhance the search algorithm's effectiveness by drawing on their expertise in information retrieval and database navigation.

Prompt 3.4 (Advanced Search Algorithm)
As an experienced librarian specialising in literature reviews and information retrieval from scholarly databases such as PhilPapers and JSTOR, your task is to design a comprehensive, precise, and customised search algorithm for systematically identifying and retrieving relevant articles related to the research question: "What is the essence of gauge structure in quantum field theory?"

Steps to follow:

1. **Decompose the Research Question:** Break down the research question into key concepts. For example, in this case, the main concepts are "gauge structure," "quantum field theory," and "philosophy of science."
2. **Identify Keywords and Subject Terms:** For each key concept, list synonyms and related terms commonly used in the philosophy of science. For instance:
    - *Gauge Structure:* gauge theory, gauge symmetry, symmetry breaking.
    - *Quantum Field Theory:* QFT, quantum fields, particle physics.
    - *Philosophy of Science:* philosophy of physics, metaphysics, epistemology.
3. **Apply Boolean Operators for Expansion and Refinement:**
    - Use the **OR** operator to combine synonyms and related terms for each concept, expanding the search to include different expressions of the same idea.
    - Use the **AND** operator to combine different key concepts, ensuring that search results focus on articles addressing all relevant aspects of the research question.
4. **Construct the Search String:** Assemble the keywords and operators into a coherent search string compatible with the syntax of the PhilPapers database. For example:
    "gauge symmetry" OR "gauge theory" AND "quantum field theory" AND "philosophy of physics" OR "metaphysics"
5. **Iterate and Refine:** Plan for an iterative process of testing and refining the search algorithm based on preliminary search results. Evaluate the precision and recall of the results, adjusting the keywords or Boolean operators to improve accuracy and ensure comprehensive coverage of the topic.

The LLM provided the following response:

To create a robust search algorithm for the research question "What is the essence of gauge structure in quantum field theory?" in the PhilPapers database, we proceed as follows:

Step 1: Decompose the Research Question

Key Concepts:

1. Gauge Structure
2. Quantum Field Theory
3. Philosophical Analysis

Step 2: Identify Keywords and Subject Terms

1. Gauge Structure
    - "gauge structure"
    - "gauge theory"
    - "gauge symmetry"
    - "gauge invariance"
    - "symmetry principles"
2. Quantum Field Theory
    - "quantum field theory"
    - "QFT"
    - "quantum physics"
    - "particle physics"
    - "standard model"
3. Philosophical Analysis
    - "philosophical foundations"
    - "ontology"
    - "metaphysics"
    - "epistemology"
    - "philosophy of science"
    - "theoretical explanation"

Step 3: Apply Boolean Operators

Within Concepts (Expansion using OR):

- **Gauge Structure:**
    - ("gauge structure" OR "gauge theory" OR "gauge symmetry" OR "gauge invariance" OR ""symmetry principles")
- **Quantum Field Theory:**
    - ("quantum field theory" OR "QFT" OR "quantum physics" OR "particle physics" OR "standard model")

- **Philosophical Analysis:**
  - ("philosophical foundations" OR "ontology" OR "metaphysics" OR "epistemology" OR "philosophy of science" OR "theoretical explanation")**Between Concepts (Refinement using AND):**

Combine the expanded terms of each concept using the **AND** operator:

```
("gauge structure" OR "gauge theory" OR "gauge symmetry" OR "gauge invariance" OR "symmetry principles") AND ("quantum field theory" OR "QFT" OR "quantum physics" OR "particle physics" OR "standard model") AND ("philosophical foundations" OR "ontology" OR "metaphysics" OR "epistemology" OR "philosophy of science" OR "theoretical explanation")
```

Step 4: Construct the Search String

PhilPapers supports standard Boolean operators and phrase searches. The constructed search string would look like this:

Final Search String:

```
("gauge structure" OR "gauge theory" OR "gauge symmetry" OR "gauge invariance" OR "symmetry principles") AND ("quantum field theory" OR "QFT" OR "quantum physics" OR "particle physics" OR "standard model") AND ("philosophical foundations" OR "ontology" OR "metaphysics" OR "epistemology" OR "philosophy of science" OR "theoretical explanation")
```

Additional Filters:

- **Publication Date:** Consider filtering by publication date to include recent developments, e.g., from 2000 to the present.
- **Language:** Filter for articles in English to ensure comprehensibility.
- **Document Type:** Focus on peer-reviewed journal articles and books for scholarly reliability.

Step 5: Iterate and Refine

- **Test the Search:** Run the initial search and evaluate the relevance and volume of the results.
- **Analyze Results:** Identify any irrelevant or missing key literature.
- **Adjust Terms:** Modify the keywords by adding or removing terms based on initial findings. For example, if too many irrelevant results pertain to "particle physics" in a non-philosophical context, consider removing or replacing that term.

- **Re-run and Evaluate:** Continue this iterative process until the search results consistently yield highly relevant literature.**Incorporate Key Authors:**

  To further refine the search, include known scholars in the field:

- **Authors to Consider:**
  - James Weatherall
  - Tim Maudlin
  - Michael Redhead
  - Katherine Brading
  - Richard Healey**Author-Specific Search String Example:**

```
(("gauge structure" OR "gauge theory" OR "gauge symmetry") AND ("quantum field theory" OR "QFT") AND ("philosophy of science")) AND (author:"James Weatherall" OR author:"Tim Maudlin" OR author:"Michael Redhead" OR author:"Katherine Brading" OR author:"Richard Healey")
```

This structured approach guarantees a comprehensive and focused literature search, allowing researchers to capture the various dimensions of the research question. By systematically breaking down the inquiry, expanding and refining search terms, and continuously testing and adjusting the search algorithm, researchers can effectively navigate extensive resources in databases like PhilPapers, ensuring that the most relevant scholarly works are identified.

4 Literature Screening and Quality Assessment

As we move from executing our search algorithm to compiling a collection of literature relevant to the research question—"What is the essence of gauge structure in quantum field theory?"—it is crucial to critically assess the quality and relevance of the materials for inclusion in our review. To ensure a rigorous and effective screening process, particularly when using AI tools, it is important to establish clear selection criteria in advance.

4.1 Quality Assessment Criteria

A set of positive and negative criteria will guide the evaluation of the literature, ensuring that the selected works contribute meaningfully to the research question.

4.1.1 Positive Criteria for Inclusion

The most commonly considered positive criteria for inclusion typically include the following aspects.

1. **Subject Matter:** The literature must explicitly address gauge structures, particularly within classical or quantum field theories. This ensures direct relevance to the research question.
2. **Philosophical Underpinnings:** Selected works should be based on well-established philosophical frameworks, such as realism, instrumentalism, or structuralism, that deepen our understanding of the physical world and its theoretical representations.
3. **Academic Rigor:** The literature must consist of scholarly works, such as original research articles published in peer-reviewed journals or validated pre-prints, ensuring that they meet high academic standards.
4. **Linguistic Consistency:** To maintain consistency and allow for thorough analysis, the literature should be published in English. This criterion ensures that all selected works can be seamlessly integrated into the review.

By adhering to these criteria, we aim to curate a body of literature that is not only directly relevant to the research question but also meets standards of academic rigor, philosophical depth, and accessibility. This approach will help build a strong intellectual foundation for the literature review.

4.1.2 Negative Criteria for Exclusion

In addition to positive selection criteria, it is essential to define exclusion criteria to filter out less relevant or inappropriate materials. This will help maintain the academic integrity and focus of the literature review.

1. **Language Exclusion:** Articles not published in English will be excluded to ensure that all selected works are accessible and can be fully analysed within the scope of the study.
2. **Originality:** Non-original works, such as review papers, editorials, or commentaries, will generally be excluded unless they provide significant new insights or perspectives. This criterion prioritises primary research that contributes directly to the discourse on gauge structures in quantum field theory.
3. **Irrelevance:** Literature that does not specifically address gauge structures within the context of quantum field theory will be excluded, even if it pertains to broader topics within physics or the philosophy of science. This helps maintain a focused inquiry.
4. **Methodological Rigor:** Studies lacking a clear methodological framework or those not meeting peer review standards will be excluded to uphold the methodological integrity of the review.

5. **Publication Status:** Works that have not been formally published or are still in the pre-print stage without subsequent peer review will be excluded to ensure that the literature has undergone the standard academic validation process.

By applying these exclusion criteria alongside the positive selection standards, we can ensure that the literature review remains focused, methodologically sound, and highly relevant to the investigation of gauge structures in quantum field theory.

This dual approach of positive and negative criteria will refine the focus of the review, ensuring it provides a concise and authoritative examination of the subject. Through this systematic process, AI tools can be guided to assist in screening papers based on these criteria, ultimately enhancing the quality and relevance of the literature review.

4.2 *Teaching the LLM to Apply the Quality Assessment Criteria*

To effectively utilise AI tools in the literature review process, it is essential to teach the LLM how to apply the established quality assessment criteria. This ensures that the LLM assists in selecting high-quality, relevant literature that aligns with the research objectives and maintains academic rigour.

Prompt 4.2.1 (Quality Assessment Criteria)
Based on the discussions above, please establish a positive and negative literature assessment framework for the research topic, "What is the essence of gauge structure in quantum field theory?"

The LLM provided the following response:

Drawing on the criteria discussed, here is a structured framework for the positive and negative assessment of literature related to the research theme, "What is the essence of gauge structure in quantum field theory?"

Positive Assessment Framework (Inclusion Criteria)

1. **Relevance to Gauge Theory:** The literature must directly address gauge structures, particularly in the context of quantum field theory. This ensures that the content is pertinent to the research question.
2. **Philosophical Depth:** The works should contribute meaningfully to the philosophical discourse, engaging with topics such as realism, instrumentalism, and the nature of scientific explanation as they relate to gauge structures.
3. **Original Research:** The selected articles should present original research findings, theories, or models that advance understanding in the field.
4. **Peer-Reviewed:** The literature must have undergone peer review, ensuring a level of academic scrutiny and validation.

5. **English Language:** To maintain consistency and accessibility, the works must be published in English.
6. **Methodological Rigour:** Studies should demonstrate a clear and robust methodology that is appropriate to the research question, ensuring reliability and validity.
7. **Citation Impact:** Preference may be given to highly cited works, indicating their influence and acceptance within the academic community.
8. **Current Trends:** The literature should reflect current developments in the philosophy of physics and quantum field theory, ensuring that the review remains up-to-date.

Negative Assessment Framework (Exclusion Criteria)

1. **Irrelevant Topics:** Exclude works that do not address the core themes of gauge structures or quantum field theory.
2. **Non-Original Content:** Exclude editorials, opinion pieces, and review papers that do not offer new insights or original data.
3. **Non-Peer-Reviewed:** Exclude documents that have not been subject to peer review, such as pre-prints or conference abstracts that lack validation.
4. **Language Barrier:** Exclude articles not published in English to ensure consistency and facilitate comprehensive analysis.
5. **Lack of Methodological Clarity:** Exclude studies that do not clearly outline their methodology or that lack sufficient methodological rigour.
6. **Insufficient Philosophical Engagement:** Exclude works that do not provide adequate philosophical depth or fail to engage critically with existing literature.
7. **Outdated Research:** While historical context is valuable, exclude outdated works that no longer contribute meaningfully to current discussions in the field.
8. **Low Citation Impact:** Exclude papers with minimal citations, as this may indicate limited academic influence or relevance.

As we can see, the LLM has well-accepted the positive and negative criteria framework that we have established. This framework ensures a balanced approach to literature selection, emphasizing the importance of both including high-quality, relevant works and excluding those that do not meet the established criteria. This approach will support a comprehensive and focused literature review that contributes meaningfully to the understanding of gauge structures in quantum field theory.

4.3 Literature Screening

After establishing the quality assessment criteria, the next crucial step is to screen the literature to determine which works should be included in the review. This process involves evaluating each piece of literature against the positive and negative criteria, ensuring that only the most relevant and high-quality sources are selected for inclusion.

Prompt 4.3.1 (Literature Screening)
We then have a literature screening case based on the literature assessment framework we just obtained. Please understand these two cases; next, you will help me screen more relevant papers.

Title 1: *Understanding Gauge.*

Abstract 1: I consider two usages of the expression "gauge theory". On one hand, a gauge theory is a theory with excess structure; on the other hand, a gauge theory is any theory appropriately related to classical electromagnetism. I make precise one sense in which one formulation of electromagnetism, the paradigmatic gauge theory on both usages, may be understood to have excess structure, and then argue that gauge theories on the second usage, including Yang-Mills theory and general relativity, do not generally have excess structure in this sense.

Screening Result: Accepted.

Title 2: *Categories and the Foundations of Classical Field Theories.*

Abstract 2: I review some recent work on applications of category theory to questions concerning theoretical structure and theoretical equivalence of classical field theories, including Newtonian gravitation, general relativity, and Yang-Mills theories.

Screening Result: Rejected;

Reason: Not an original article, but a summary of research findings.

When we uploaded the PDF file from https://arxiv.org/abs/1412.0303, the LLM provided the following assessment:

Assessment:

1. **Relevance:** The article is relevant to general relativity, which is connected to the broader theme of gauge structures within classical field theories.
2. **Originality:** The abstract suggests that the paper presents a novel perspective or result, which is a positive factor for inclusion.
3. **Methodology:** It appears that the paper offers a new formulation or clarification, indicating a methodological contribution.
4. **Language:** Assuming the paper is in English, it meets the language criterion.
5. **Peer Review:** Although peer review status is not explicitly mentioned, it is generally expected that articles on such topics are peer-reviewed, which aligns with the inclusion criteria.

Given the abstract, the article seems to meet several positive criteria for inclusion. However, the final decision would depend on its alignment with the specific research question and whether it provides philosophical insights into gauge structures. A more detailed examination of the paper would be required to assess its suitability fully.

Screening Result: *Tentatively accepted*, pending confirmation of peer review status and further evaluation of its alignment with the philosophical aspects of the research question.

5 Data Extraction and Standardization

Data extraction and standardisation represent a critical phase in the literature review process, providing the foundation for a coherent and comprehensive analysis. This section focuses on systematically extracting and documenting key information from the selected literature. Important elements to capture include authorship, publication details, central theses, and the methodologies used in the arguments. By standardising this information, researchers can compare and synthesise various philosophical perspectives more effectively, contributing to a robust and insightful literature review.

Standardisation is essential for ensuring consistency and comparability across diverse sources. It enables the alignment of different arguments and methodologies, making it easier to identify patterns, contradictions, and gaps in the literature. This process not only enhances the organisation of the review but also strengthens the analytical framework, leading to a more thorough synthesis of the reviewed material. In doing so, standardisation supports the methodological rigour of the literature review while contributing to the overall depth and quality of the final analysis.

5.1 Defining Standard Information

For our research on "What is the essence of gauge structure in quantum field theory?", it is essential to define the standard information that should be extracted from each relevant reference. Leveraging LLMs can assist in generating a consistent list of this standard information. Below is a structured process for achieving this:

> **Prompt 5.1.1 (Standard Information Clarification)**
> Considering the ongoing research topic, "What is the essence of gauge structure in quantum field theory?", what standard information should we extract from each related reference?

LLM's suggested standard information:
Authorship and Publication Details:

- **Author(s):** Identify the primary researchers or theorists and their institutional affiliations.
- **Publication Date and Venue:** Record the date and location of publication, including the journal or conference.
- **Title:** Ensure the title reflects a clear focus on gauge structures within quantum field theory.

Central Thesis:

- **Main Argument:** Summarise the paper's core thesis or argument. What is the author's perspective on the essence of gauge structure?
- **Objective:** Clarify whether the study aims to propose a new interpretation, critique existing theories, or explore specific aspects of gauge structure.

Methodologies:

- **Theoretical Framework:** Identify the theoretical approaches employed, such as specific quantum field theories, gauge theories, or mathematical frameworks (e.g., fibre bundles, Lie groups).
- **Analytical Methods:** Document any specific analytical methods, such as symmetry analysis, topological considerations, or mathematical formalism.

Key Concepts and Definitions:

- **Gauge Structure:** Note how the paper defines and conceptualises gauge structure within quantum field theory.
- **Related Concepts:** Record other significant concepts discussed, such as gauge invariance, symmetry breaking, or the role of gauge fields.

Findings and Contributions:

- **Core Findings:** Highlight the key findings regarding the essence of gauge structures. Does the paper present a novel interpretation or challenge existing views?
- **Implications:** Consider the broader implications of these findings for quantum field theory and related fields.

Citations and Influences:

- **Referenced Works:** Note foundational works or theories that the paper references, particularly those related to the development of gauge theories.
- **Influence on the Field:** Assess how this paper has contributed to ongoing debates or developments in the understanding of gauge structures.

Critical Evaluation:

- **Strengths and Weaknesses:** Evaluate the strengths and weaknesses of the paper's arguments and methodologies.
- **Relevance:** Determine the paper's relevance to the research question. Does it directly address the essence of gauge structure, or does it provide necessary background or context?

By systematically collecting and standardising this information, researchers can build a comprehensive understanding of how different scholars have approached the concept of gauge structures in quantum field theory. This structured approach will significantly aid in synthesising these perspectives in the literature review, ultimately leading to a well-rounded and insightful analysis.

5.2 Data Extraction and Standardisation

After defining the key elements to be extracted from the literature, the next step is to systematically collect and present this information in a standardised format. This process is essential for ensuring consistency across different sources, making comparisons and synthesis more straightforward and effective.

In this subsection, we will demonstrate how to apply the established criteria to extract relevant data from selected papers. Organising the extracted information in a structured format, such as a table, facilitates clear and consistent documentation. This method ensures that each source contributes meaningfully to the overall analysis of the research question: "What is the essence of gauge structure in quantum field theory?"

Prompt 5.2.1 (Data Extraction and Standardisation)
Please extract the relevant elements from the paper I uploaded based on the previously defined standard information and present the results in a tabular format.

The LLM generates the following data extraction table.

Based on the content extracted from the uploaded document, here's a table summarizing the standard information relevant to your research topic, "What is the essence of gauge structure in quantum field theory?"

Category	Details
Authorship	James Owen Weatherall
Publication Details	Department of Logic and Philosophy of Science, University of California, Irvine, CA 92697; Arxiv ID: 1505.02229v2, Date: 13 December 2017
Central Thesis	The paper discusses two primary uses of "gauge theory": one involving excess structure, and the other involving theories related to classical electromagnetism.
Methodologies	The paper contrasts different formulations of electromagnetism (EM1 and EM2) and their relationship to gauge structure using mathematical formalism and category theory.
Key Concepts and Definitions	Gauge theory, excess structure, Yang-Mills theory, general relativity, principal bundles.
Findings and Contributions	The paper argues that while electromagnetism (EM2) can be considered to have excess structure, other theories like Yang-Mills and general relativity do not.
Citations and Influences	The paper references key works by Earman, Norton, Healey, Rosenstock, and others on gauge theory and its philosophical implications.

This table summarises the essential information from the document, providing relevant insights for your literature review on the essence of gauge structure in quantum field theory.

5.3 *Synthesising and Presenting Extracted Data*

Extracting and standardising data from various sources is just the first step in constructing a comprehensive literature review. The next crucial phase is synthesising this information to create a coherent narrative that aligns with and supports your research objectives.

When integrating the extracted information into your literature review, it is essential to present the findings in a way that clearly communicates the relevance and contribution of each source. For example, when discussing a specific paper, you should succinctly summarise the key insights it offers and connect them to the broader themes of your review.

A well-crafted sentence in the final literature review might appear as follows:

> Weatherall's examination of gauge theory offers a nuanced distinction between excess structure and its application in classical electromagnetism, arguing that while electromagnetism may exhibit such excess, other theories, such as Yang-Mills and general relativity, do not share this characteristic (Weatherall, 2017).

This sentence effectively integrates the standardised data into a concise summary, highlighting the paper's contribution to the research question. It also situates the work within the broader discourse, making its relevance to the study of the essence of gauge structure in quantum field theory clear.

By applying this approach consistently to each source, you can develop a literature review that is both comprehensive and analytical. This method ensures that your review not only covers relevant literature but also presents it in a manner that is organised, insightful, and engaging. Ultimately, this approach lays a strong foundation for further research and discussion, positioning your review as a well-rounded examination of the topic.

6 Literature Synthesis and Meta-analysis

The meta-analysis phase of the literature review is crucial for synthesising and critically evaluating the body of knowledge gathered from various sources. This stage involves systematically analysing the literature to uncover convergences and divergences among different philosophical theories, particularly in relation to the research topic. While traditional meta-analysis, typically quantitative, is less common in philosophy due to its qualitative nature, thematic or conceptual synthesis is highly effective. By identifying and analysing recurring themes, concepts, and arguments, this process enables a deeper understanding of broader trends, debates, and implications within the field.

This synthesis not only highlights areas of consensus but also illuminates points of contention, providing a comprehensive overview that guides further inquiry and analysis. A well-conducted meta-analysis can reveal the underlying structure of the discourse, helping to map the intellectual landscape of the research area.

To perform a thorough synthesis and meta-analysis of the literature, the following structured approach can be used when analysing multiple papers:

Prompt 6.1 (Meta-analysis of a Documents Batch)
Please extract and summarise the key thematic elements from each of the uploaded papers, focusing on the following aspects:

1. **Core Arguments and Theses:** Identify the central arguments or theses presented in each paper.
2. **Philosophical Perspectives:** Note the philosophical perspectives or schools of thought each paper aligns with or critiques.
3. **Methodological Approaches:** Outline the methodologies or approaches used by the authors to support their arguments.
4. **Points of Convergence:** Highlight areas where the papers converge or agree on certain concepts or arguments.
5. **Points of Divergence:** Identify where the papers diverge or present differing viewpoints or critiques.
6. **Implications for the Field:** Discuss the implications of these convergences and divergences for the broader philosophical discourse related to the research topic, "What is the essence of gauge structure in quantum field theory?"
7. **Broader Trends and Debates:** Provide a summary of the broader trends or debates that emerge from synthesising these papers.

Example of LLM's Response:

The paper explores two primary interpretations of "gauge theory." The first considers gauge theory as a framework that inherently includes excess structure. The second views gauge theory as any theory appropriately related to classical electromagnetism. The paper argues that while electromagnetism can be characterised by excess structure, other theories, such as Yang-Mills theory and general relativity, do not share this characteristic, based on the provided definition of excess structure.

Philosophically, the paper challenges the assumption that all gauge theories inherently possess excess structure. It engages with the broader philosophical discourse on the nature of gauge theories, focusing on their role and interpretation within classical field theory and general relativity. This critique contributes to ongoing debates about how gauge theories relate to the physical world and the mathematical structures that describe them.

Methodologically, the author uses a comparative analysis of different formulations of electromagnetism to explore these concepts. Through mathematical formalism and category theory, the paper articulates the notion of excess structure and contrasts it across various theoretical frameworks. When more related literature is provided, the LLM could perform further meta-analysis:

Synthesis and Broader Implications
There is a general convergence in the literature regarding the understanding that not all gauge theories possess excess structure. This consensus is particularly evident when comparing different theories such as electromagnetism and Yang-Mills theory. The recognition of fundamental differences in the structural properties of these theories represents an important point of agreement among scholars.

However, significant divergence exists regarding what constitutes excess structure in gauge theories. Some scholars view this excess as philosophically significant, while others consider it a mathematical artefact. This divergence reflects varying opinions on the role and importance of excess structure within gauge theories.

The paper suggests a more nuanced understanding of gauge structures in quantum field theory, challenging the assumption that all gauge theories share the same foundational structure. This perspective has significant implications for how these theories are interpreted, both in physics and in philosophical discourse. It calls for a reconsideration of the fundamental principles underlying gauge theories and encourages a more critical examination of their structural aspects.

A broader trend in the literature reflects a shift towards differentiating between various types of gauge theories based on their structural properties. This paper contributes to the ongoing debate about whether excess structure is a necessary component of gauge theories or a redundant feature. The discussion adds depth to the conversation, pushing the field towards a more refined understanding of the complexities involved in gauge theory.

7 Conclusion

In this comprehensive exploration of literature review methodologies, we have highlighted the key stages in researching the essence of gauge structures in quantum field theory, a topic bridging physics and philosophy. The work began with a well-defined research question and careful selection of bibliographic databases, followed by the development of a robust search algorithm tailored to retrieve high-quality, relevant literature. Systematic screening and quality assessment ensured that only the most pertinent studies were included. By extracting and standardising key elements such as authorship, central theses, and methodologies, we facilitated meaningful comparisons and synthesis across different philosophical perspectives. Meta-analysis further illuminated points of convergence and divergence in the literature, providing a nuanced understanding of the debates surrounding gauge structures, while frontier mapping identified emerging trends and suggested new research avenues. A significant value of this review lies in its demonstration of how AI enhances both the efficiency and depth of the research process. From generating search algorithms to extracting and summarising information, AI has proven invaluable. However, human judgment remains essential in maintaining the quality and relevance of the final review. Looking ahead, as AI technologies continue to evolve, they offer the potential to significantly advance scholarly research, enabling deeper analysis, the identification of complex patterns, and the prediction of future trends. Ultimately, this work not only deepens the understanding of gauge structures but also illustrates the powerful synergy between AI and human expertise in shaping the future of academic inquiry.

Acknowledgements This research was supported by the Ministry of Education Humanities and Social Science Research Special Project, "A Study on Mechanisms for Promoting High-Quality Employment of University Graduates from the Perspective of Fostering New Productive Forces in Universities" (Project No. 24JDSZ3170). The author gratefully acknowledges this support, which has contributed to the successful completion of this stage of the research.

References

Anderson, L., Krathwohl, D., Airasian, P., Cruikshank, K., Mayer, R., Pintrich, P., Raths, J., & Wittrock, M. (Eds.). (2000). *A taxonomy for learning, teaching, and assessing: A revision of Bloom's taxonomy of educational objectives*. Pearson.

Angrist, J. D., & Pischke, J. S. (2009). *Mostly harmless econometrics: An empiricist's companion*. Princeton University Press.

Baumeister, R. F., & Leary, M. R. (1997). Writing narrative literature reviews. *Review of General Psychology, 1*(3), 311–320. https://doi.org/10.1037/1089-2680.1.3.311

Boote, D. N., & Beile, P. (2005). Scholars before researchers: On the centrality of the dissertation literature review in research preparation. *Educational Researcher, 34*(6), 3–15. https://doi.org/10.3102/0013189X034006003

Floridi, L., & Cowls, J. (2019). A unified framework of five principles for AI in society. *Harvard Data Science Review, 1*(1), 1–15. https://doi.org/10.1162/99608f92.8cd550d1

Glass, G. V. (1976). Primary, secondary, and meta-analysis of research. *Educational Researcher, 5*(10), 3–8. https://doi.org/10.3102/0013189X005010003

Grant, M. J., & Booth, A. (2009). A typology of reviews: An analysis of 14 review types and associated methodologies. *Health Information & Libraries Journal, 26*(2), 91–108. https://doi.org/10.1111/j.1471-1842.2009.00848.x

Hart, C. (1998). *Doing a literature review: Releasing the social science research imagination*. Sage.

Higgins, J. P. T., & Green, S. (Eds.). (2011). *Cochrane handbook for systematic reviews of interventions* (Version 5.1.0). The Cochrane Collaboration. Retrieved from https://handbook-5-1.cochrane.org

Liberati, A., Altman, D. G., Tetzlaff, J., Mulrow, C., Gøtzsche, P. C., Ioannidis, J. P. A., Clarke, M., Devereaux, P. J., Kleijnen, J., & Moher, D. (2009). The PRISMA statement for reporting systematic reviews and meta-analyses of studies that evaluate healthcare interventions: Explanation and elaboration. *PLoS Medicine, 6*(7), e1000100. https://doi.org/10.1371/journal.pmed.1000100

Machi, L. A., & McEvoy, B. T. (2016). *The literature review: Six steps to success* (3rd ed.). Corwin Press.

Moher, D., Liberati, A., Tetzlaff, J., Altman, D. G., & The PRISMA Group. (2009). Preferred reporting items for systematic reviews and meta-analyses: The PRISMA statement. *PLoS Medicine, 6*(7), e1000097. https://doi.org/10.1371/journal.pmed.1000097

Ridley, D. (2012). *The literature review: A step-by-step guide for students* (2nd ed.). Sage.

Torraco, R. J. (2005). Writing integrative literature reviews: Guidelines and examples. *Human Resource Development Review, 4*(3), 356–367. https://doi.org/10.1177/1534484305278283

Chapter 5
Using AI for Data Analysis

Zhenyu Sun and Ruixue Yuan

Abstract Artificial Intelligence (AI) has become a cornerstone of modern data analysis, revolutionising how data is interpreted, modelled, and visualised. This chapter examines the role of AI in both quantitative and qualitative data analysis, with a focus on statistical analysis, predictive modelling, and advanced data visualisation techniques. AI's machine learning and deep learning frameworks enhance the detection of complex patterns in large datasets, enabling more accurate predictions and deeper insights. The chapter also explores AI-driven tools for handling high-dimensional data, automating statistical methods such as Analysis of Variance (ANOVA) and regression analysis. In addition, generative AI (GenAI), especially large language models like ChatGPT, has been utilised to assist in data processing and analysis, providing innovative methods for handling both structured and unstructured data. By examining these developments, this chapter highlights the significant contributions of AI in data analysis, pointing towards future innovations in research methodologies.

1 Introduction

Artificial Intelligence (AI) has become a transformative force in data analysis, reshaping the methodologies used to interpret, analyse, and apply data across various disciplines. AI incorporates technologies such as machine learning, deep learning, natural language processing, and advanced statistical techniques, each of which contributes to enhancing the data analysis process by enabling the efficient handling of large datasets with unprecedented speed and precision. More importantly, AI

Z. Sun
University of Chinese Academy of Sciences, Beijing, China

R. Yuan (✉)
Beijing University of Chemical Technology, Beijing, China
e-mail: yuanruixue@buct.edu.cn

X. Zhou, H. Al-Samarraie (eds.), *Institutional guide to using AI for research*, Advances in Artificial Intelligence in Education 2,
https://doi.org/10.1007/978-3-031-94809-1_5

excels in identifying complex patterns, correlations, and trends that traditional analytical methods often miss.

The integration of AI in data analysis has allowed researchers and practitioners to move beyond conventional methods, providing innovative solutions to the challenges posed by increasingly complex and high-dimensional datasets. AI-driven models and algorithms not only streamline analytical processes but also make them more efficient, accurate, and insightful. This has led to discoveries and conclusions previously out of reach, underscoring AI's profound impact on modern data practices (Subasi, 2020).

As the volume, variety, and velocity of data continue to grow, traditional methods often fail to address the complexity of contemporary datasets. AI addresses these challenges by offering scalable and adaptive solutions, capable of managing the immense scale and intricate nature of modern data (Rahmani et al., 2021). In addition to enhancing efficiency, AI automates routine analytical tasks, enabling researchers to concentrate on more strategic and interpretive aspects of their work. This shift boosts productivity and enables the extraction of deeper insights, fostering innovation across various sectors. In areas such as healthcare, finance, and social sciences, AI-driven data analysis has led to breakthroughs that significantly advance both understanding and practical applications (Patil & Shankar, 2023; George et al., 2023).

AI represents not just a tool but a paradigm shift in how data is approached and utilised. Its ability to process, analyse, and visualise data in ways previously unimaginable has made it an indispensable component of modern scientific research and industrial applications. This chapter explores AI's role in various aspects of data analysis, including statistical evaluation, predictive modelling, and data visualisation, emphasising its transformative impact on contemporary research methodologies. Specific applications and methods where AI's influence is most pronounced are examined, providing a comprehensive overview of its critical role in the evolving landscape of data analysis.

2 AI in Quantitative Data Analysis

2.1 Statistical Analysis

AI has revolutionized quantitative data analysis by enhancing traditional statistical methods, leading to more efficient, accurate, and scalable approaches for managing complex datasets. In this section, we explore two fundamental statistical techniques—ANOVA and regression analysis—illustrating how AI can automate and improve these processes through detailed examples.

2.1.1 Automated Analysis of Variance (ANOVA)

To demonstrate the power of AI in automating statistical analysis, consider a scenario involving the comparison of means across multiple groups. This example highlights how AI can streamline and enhance the traditional ANOVA process, making it more efficient and insightful.

Comparing the Means of Multiple Groups Using Automated ANOVA
Imagine a researcher investigating the impact of three different diets—Diet A, Diet B, and Diet C—on weight loss among a group of individuals. The goal is to determine whether there are statistically significant differences in the mean weight loss across these diets.

Step 1: Data Preparation

The first step in data preparation involves structuring the dataset, which includes weight loss measurements from participants divided into three groups, each corresponding to a different diet.

The dataset contains two key variables: Group and Weight_Loss. The Group variable is categorical, representing the diet group each participant followed, labelled as A, B, or C. The Weight_Loss variable is continuous and captures the amount of weight lost by each participant. The data is organised in a table format where each row represents an individual participant's results.

For example, the dataset can be created in Python using pandas:

```
import pandas as pd
# Example dataset
data = {
    'Group': ['A', 'A', 'A', 'B', 'B', 'B', 'C', 'C', 'C'],
    'Weight_Loss': [4.5, 5.0, 3.5, 2.5, 3.0, 2.0, 6.0, 5.5, 6.5]
}

df = pd.DataFrame(data)
```

In this structure, each row shows the diet group and the corresponding weight loss for a particular participant. This format provides the necessary data for further analysis, such as performing statistical tests to compare weight loss across the different diet groups.

Step 2: Applying Automated ANOVA Using AI

Traditionally, conducting ANOVA requires manual calculation of the F-statistic or using statistical software. However, AI can automate this process, making it more efficient and reducing the likelihood of human error. The following code uses the statsmodels library to apply ANOVA:

```
from statsmodels.formula.api import ols
from statsmodels.stats.anova import anova_lm
# Fit the model
model = ols('Weight_Loss ~ Group', data=df).fit()

# Perform ANOVA
anova_results = anova_lm(model)
print(anova_results)
```

This model uses Ordinary Least Squares (OLS) to fit the data, and ANOVA is performed to compare the means across the diet groups. The anova_lm function automatically computes the F-statistic and p-value, which are critical for determining whether the differences between groups are statistically significant.

Step 3: Interpreting the Results

The ANOVA results provide the F-statistic and p-value, which help interpret whether the differences between the group means are significant. If the p-value is less than a predetermined significance level (e.g., 0.05), the null hypothesis (that all group means are equal) is rejected.

For example, consider the following ANOVA output:

```
df    sum_sq   mean_sq   F     PR(>F)
Group         2    20.25     10.125   15.19  0.003
Residual      6    4.00      0.667
```

In this case, the F-statistic is 15.19, with a p-value of 0.003, indicating a significant difference in weight loss among the diets.

Step 4: Enhancements Using AI

AI enhances the traditional ANOVA process by suggesting additional analyses and visualisations. For example, if ANOVA indicates significant differences, AI can automatically recommend and perform post-hoc tests, such as Tukey's Honestly Significant Difference (HSD) test, to identify which groups differ from each other.

```
from stats models.stats.multicomp import pairwise_tukeyhsd
# Perform Tukey's HSD test
tukey_results = pairwise_tukeyhsd(endog=df['Weight_Loss'],
groups=df['Group'], alpha=0.05)
print(tukey_results)
```

The Tukey HSD results provide pairwise comparisons between groups, pinpointing where the significant differences lie. Additionally, AI can generate visualisations, such as box plots or interaction plots, to provide intuitive insights into the data.

By automating the ANOVA process, AI reduces manual calculation errors and enables researchers to perform comprehensive analyses more efficiently. This integration of AI not only simplifies the process but also adds depth to the analysis, making the results more interpretable and robust. AI-driven automation is particularly valuable when dealing with large datasets or complex experimental designs, ensuring that statistical analyses are both rigorous and accessible.

In the final subsection of this section, we will demonstrate how generative AI can assist in executing the tasks outlined in this chapter.

2.1.2 Regression Analysis and AI Enhancements

Regression analysis is another foundational statistical technique used to predict a dependent variable based on one or more independent variables. AI significantly enhances traditional regression methods by improving the accuracy and robustness of predictions. In this subsection, we explore AI's impact through a detailed example of predicting house prices.

Predicting House Prices Using AI-Enhanced Regression Analysis
Consider a real estate analyst tasked with predicting house prices based on factors such as square footage, the number of bedrooms, and location. While traditional linear regression can provide a basic predictive model, AI techniques capture more complex relationships, improving the accuracy of the model.

Step 1: Data Preparation

The first step in regression analysis is preparing the dataset, which includes several predictors, or independent variables, that influence house prices.

These predictors consist of Square_Feet, a continuous variable representing the size of the house, and Bedrooms, a categorical variable indicating the number of bedrooms. Additionally, Location is another categorical variable that represents the neighbourhood, with categories such as Suburb, Urban, and Rural. The target variable, Price, is the dependent variable representing the house price.

To illustrate this, we can create a simple dataset using Python and pandas:

```
import pandas as pd


# Example dataset
data = {
    'Square_Feet': [1500, 2000, 2500, 1800, 1200],
    'Bedrooms': [3, 4, 4, 3, 2],
    'Location': ['Suburb', 'Suburb', 'Urban', 'Urban', 'Rural'],
    'Price': [300000, 400000, 450000, 350000, 200000]
}

df = pd.DataFrame(data)
```

In this dataset, Square_Feet and Bedrooms serve as the key predictors, while Price is the variable we aim to predict. Since Location is categorical, it may need to be encoded into numerical values if included in the regression model. This structured data provides the necessary foundation for conducting the regression analysis.

Step 2: Applying Traditional Regression Analysis

Once the data is prepared, we can apply traditional linear regression to predict house prices. Linear regression models the relationship between the predictors and the target variable as a linear function:

$$\text{Price} = \beta_0 + \beta_1 \times \text{Square\_Feet} + \beta_2 \times \text{Bedrooms} + \epsilon$$

where β_0 is the intercept, β_1 and β_2 are the coefficients for the predictors, and ϵ is the error term. Here's how you might implement this using Python:

```
from sklearn.linear_model import LinearRegression


# Prepare the data
X = df[['Square_Feet', 'Bedrooms']]
y = df['Price']

# Fit the model
model = LinearRegression()
model.fit(X, y)

# Make predictions
predictions = model.predict(X)
print(predictions)
```

This model fits the data and predicts house prices based on the size of the house and the number of bedrooms.

Step 3: AI Enhancements

While linear regression provides a straightforward model, it may not capture all the nuances in the data, especially non-linear relationships. AI can enhance this traditional approach by introducing more advanced techniques.

One such technique is polynomial regression, which helps model non-linear relationships by adding polynomial terms (e.g., squared terms) to the predictors. For instance, by incorporating squared terms for variables like Square_Feet and Bedrooms, the model can account for more complex relationships between the features and house prices. Below is an example using Python, where polynomial regression is applied:

```
from sklearn.preprocessing import PolynomialFeatures
from sklearn.pipeline import make_pipeline

# Polynomial regression example
poly_model = make_pipeline(PolynomialFeatures(degree=2),
LinearRegression())
poly_model.fit(X, y)
poly_predictions = poly_model.predict(X)
print(poly_predictions)
```

In this case, the addition of polynomial terms enables the model to better fit the data by accounting for non-linear interactions between the predictors and the target variable.

Furthermore, to prevent overfitting, regularisation techniques such as Lasso (L1 regularisation) or Ridge (L2 regularisation) can be employed. These methods penalise large coefficients, which helps the model generalise better to unseen data and ensures that it remains robust, as highlighted by Hastie et al. (2017).

Step 4: Model Evaluation and Interpretation

Once the model is enhanced using AI techniques, its performance needs to be evaluated. Common metrics include Mean Squared Error (MSE) and R-squared (R^2):

$$\mathrm{MSE} = \frac{1}{n}\sum_{i=1}^{n}\left(y_i - \hat{y}_i\right)^2$$

$$R^2 = 1 - \frac{\sum_{i=1}^{n}\left(y_i - \hat{y}_i\right)^2}{\sum_{i=1}^{n}\left(y_i - \overline{y}\right)^2}$$

where y_i are the actual prices, $\hat{y}_i$ are the predicted prices, and $\bar{y}$ is the mean of the observed prices. Here's how you might evaluate the enhanced model:

```
from sklearn.metrics import mean_squared_error, r2_score


# Evaluate the model
mse = mean_squared_error(y, poly_predictions)
r2 = r2_score(y, poly_predictions)
print(f"MSE: {mse}, R-squared: {r2}")
```

These metrics help assess the model's fit. A lower MSE and higher R^2 value indicate a better model fit.

By enhancing traditional regression analysis with AI techniques, the model becomes more robust and capable of capturing complex relationships within the data. Polynomial regression allows the model to understand non-linear trends, while regularisation ensures that the model generalises well to new data. This approach provides analysts with powerful tools to address real-world predictive tasks, offering insights that are both accurate and interpretable.

3 Using GenAI to Assist with Automated Statistical Analysis

The advent of generative AI tools, such as ChatGPT, has revolutionised how researchers approach statistical analysis by automating many of the more complex steps involved in data processing. These conversational agents not only simplify the execution of statistical tests like ANOVA and t-tests but also provide real-time guidance and code generation for researchers who may not have extensive programming or statistical knowledge. By making the process more accessible, GenAI tools help streamline analysis workflows, enabling researchers to focus on interpreting results and deriving insights.

3.1 Applying GenAI to Perform ANOVA and T-tests

With tools like ChatGPT, researchers can easily automate statistical analyses, such as ANOVA and t-tests, through simple, natural language queries. For example, a researcher working with weight loss data across three diet groups could prompt ChatGPT with a question like:

Prompt 3.1.1
"I have weight loss data for three groups: Diet A, Diet B, and Diet C. Can you help me prepare this data for ANOVA?"

ChatGPT would respond by generating code to structure the dataset using Python's pandas library. After the dataset is structured, the user could then ask ChatGPT to perform the ANOVA test. Rather than manually coding the analysis, the researcher can request:

> **Prompt 3.1.2**
> "Can you generate Python code to perform ANOVA on this dataset?"

ChatGPT will automatically generate the necessary code, similar to the manual approach, using libraries such as statsmodels. The generated code fits an OLS model to the data and performs ANOVA to compare the means across the diet groups. The entire process, from data preparation to result interpretation, is made easier and faster through automation.

In addition to ANOVA, ChatGPT can also assist with performing t-tests. A researcher could ask:

> **Prompt 3.1.3**
> "How do I perform a t-test between Diet A and Diet B?"

ChatGPT would generate the Python code to conduct an independent t-test, as shown below:

```
from scipy.stats import ttest_ind


# Split dataset for two groups (A and B)
group_A = df[df['Group'] == 'A']['Weight_Loss']
group_B = df[df['Group'] == 'B']['Weight_Loss']

# Perform t-test
t_stat, p_value = ttest_ind(group_A, group_B)
print(f"T-statistic: {t_stat}, P-value: {p_value}")
```

This automation allows researchers to compare two specific groups without needing to write the code from scratch, streamlining the analysis. Moreover, ChatGPT not only generates the code but can also interpret the statistical output for the researcher. For instance, after performing ANOVA, the researcher might ask:

> **Prompt 3.1.4**
> "What does an F-statistic of 15.19 and a p-value of 0.003 mean?"

ChatGPT would explain that the p-value being lower than 0.05 indicates a statistically significant difference in weight loss between the diet groups, helping researchers interpret their results with confidence.

Finally, ChatGPT can generate code for computing descriptive statistics and creating visualisations. A user might request:

Prompt 3.1.5
"Can you provide the code to compute descriptive statistics for each group?"

ChatGPT would generate the following Python code to compute key metrics such as mean, median, and standard deviation:

```
# Generate descriptive statistics
descriptive_stats = df.groupby('Group').describe()
print(descriptive_stats)
```

In addition to statistical summaries, ChatGPT can create visualisations like box plots to assist researchers in visualising their data:

```
import seaborn as sns
import matplotlib.pyplot as plt

# Create box plot
sns.boxplot(x='Group', y='Weight_Loss', data=df)
plt.title('Weight Loss by Diet Group')
plt.show()
```

Through such interactions, researchers can automate data analysis and visualisation processes without extensive coding expertise, making complex statistical analysis more accessible.

3.2 *Enhancing Accessibility and Efficiency with ChatGPT*

Generative AI tools such as ChatGPT significantly enhance the accessibility and efficiency of statistical analysis. By providing real-time code generation and guidance, these tools enable researchers to conduct rigorous analyses with minimal manual input. This is particularly beneficial for those without advanced programming or statistical backgrounds, as the tool bridges the gap between expertise and application.

For experienced researchers, ChatGPT offers an efficient way to automate routine tasks, allowing more time for in-depth analysis and interpretation. In addition to performing quantitative analyses like ANOVA and t-tests, ChatGPT's flexibility extends to qualitative data analysis. By using natural language processing (NLP) techniques, it can summarise large text datasets or identify key patterns, providing support across various research domains.

In summary, tools like ChatGPT provide significant advantages in simplifying and automating quantitative data analysis. By generating code, explaining results, and suggesting further analyses, ChatGPT ensures that statistical analysis is more user-friendly and efficient. Whether handling large datasets or managing complex experimental designs, these AI-driven tools help researchers conduct comprehensive analyses with greater ease, making them an invaluable resource in modern research workflows.

4 AI in Qualitative Data Analysis

AI has become increasingly integral in qualitative data analysis, providing researchers with advanced tools to automate and enhance the process of identifying patterns, themes, and insights from complex datasets. In this section, we explore three key methods—K-means clustering, Latent Dirichlet Allocation (LDA), and the use of NVivo in AI-enhanced qualitative research. Each method is illustrated with examples to guide students through applying these techniques to qualitative data analysis.

4.1 Use of NVivo in AI-Enhanced Qualitative Research

NVivo is a well-established qualitative data analysis software that integrates AI tools to enhance the process of coding, categorising, and analysing qualitative data. It is particularly effective for handling large volumes of text, such as interview transcripts, focus group discussions, or open-ended survey responses. In this section, we explore how NVivo can be used to analyse transcripts from focus group discussions on consumer attitudes towards sustainable products.

4.1.1 Step 1: Data Import and Initial Coding

The first step in using NVivo for qualitative analysis involves importing the text data into the software. NVivo supports a variety of file formats, allowing researchers to easily import transcripts, documents, or even audio and video files. Once the data is imported, NVivo can automatically code sections of the text based on predefined themes or emerging patterns identified by its AI tools.

For instance, the workflow starts by importing the focus group transcripts into NVivo. This can be done by dragging and dropping the files into the workspace or using the import function, which supports formats like Word documents, PDFs, and plain text files. After importing, NVivo's auto-coding feature can be applied to identify themes or concepts within the data. If predefined categories such as "Environmental Concerns" or "Product Quality" exist, NVivo can scan the transcripts and code relevant sections of text under these categories. Alternatively, its AI capabilities can detect emerging patterns and themes, allowing for a more exploratory analysis when themes are not predefined.

Here's a typical workflow:

- **Import Data**: Navigate to *Data > Import > Documents* in NVivo, and select the files containing the focus group transcripts.
- **Auto-Code Data**: Use the auto-coding wizard (*Analyze > Auto Code*) to automatically code the text. Researchers can choose to code by existing nodes (themes) or allow NVivo to suggest new nodes based on content.

4.1.2 Step 2: Enhancing Analysis with AI Tools

After the initial coding, NVivo's AI-enhanced tools can assist in deepening the analysis by helping to identify patterns, conduct sentiment analysis, and visualise relationships between themes. NVivo can automatically detect frequently occurring phrases or concepts and suggest additional coding patterns, which can uncover subtle insights that might otherwise be overlooked in manual coding.

For example, NVivo's Text Search Query tool allows researchers to search for recurring phrases or words across the dataset, making it particularly useful for identifying common concerns or attitudes among participants. Furthermore, NVivo's Sentiment Analysis feature can categorise sections of text as positive, negative, or neutral, providing valuable insights into the overall tone of the discussions and how participants feel about specific topics. NVivo's AI capabilities also include Pattern-Based Coding, where the software suggests new codes based on patterns detected in the data, such as recurring ideas or concerns not initially considered.

A typical workflow might look like this:

- **Text Search Query**: Navigate to *Explore > Queries > Text Search Query*, and input keywords related to the research (e.g., "sustainability," "quality," or "price") to find relevant sections across all transcripts.
- **Sentiment Analysis**: Use *Analyze > Sentiment Analysis* to automatically code sections of the text based on emotional tone. This helps identify which aspects of sustainable products evoke strong opinions among participants.

4.1.3 Step 3: Visualizing Results

NVivo's strength lies in its ability to visualise qualitative data, making complex findings more accessible and easier to communicate. The software offers several visualisation tools that effectively illustrate the connections between different themes, enhancing the clarity and impact of the analysis.

For instance, word clouds are a simple yet powerful way to display the most frequently occurring words in a dataset, providing a quick overview of key themes. In a word cloud, larger words indicate higher frequency, which can guide further analysis by highlighting the most prominent concepts. NVivo also offers mind maps to visualise the relationships between themes or ideas, helping to explore how various attitudes toward topics such as sustainable products are interconnected. Additionally, NVivo can generate various types of charts, including bar charts and pie charts, which represent the distribution of themes or sentiments across the dataset. These visualisations offer a clear, quantitative representation of the data, making it easier to identify trends and patterns.

The workflow for visualising results might proceed as follows:

- **Create a Word Cloud**: Go to *Explore > Word Frequency Query*, select the dataset, and generate a word cloud to see which terms are most prominent in discussions about sustainable products.
- **Generate a Mind Map**: Use *Explore > Mind Map* to visualise how different themes, such as "Environmental Impact," "Cost," and "Convenience," are related. This tool helps in understanding the broader narrative emerging from the focus groups.

NVivo significantly streamlines the qualitative research process by automating many coding and analysis tasks, allowing researchers to focus more on interpreting and developing theories. By utilising NVivo's AI-enhanced features, researchers can quickly identify patterns, sentiments, and relationships within the data, leading to richer insights and more informed conclusions. This approach not only saves time but also enhances the depth and rigour of qualitative analysis, making NVivo an invaluable tool in AI-enhanced qualitative research.

4.2 Using GenAI to Assist with Qualitative Data Analysis

While NVivo offers powerful AI-enhanced tools for qualitative data analysis, recent advancements in generative AI, such as ChatGPT, provide researchers with an even more flexible and accessible way to conduct qualitative analysis. ChatGPT allows users to automate many of the tasks traditionally performed in software like NVivo, but with the added benefit of real-time, conversational interactions that can streamline coding, pattern identification, sentiment analysis, and data visualisation. This section explores how researchers can use ChatGPT to assist with the analysis of focus group discussions on consumer attitudes towards sustainable products, providing an alternative approach to NVivo's more structured workflows.

4.2.1 Importing and Coding Text Data

The first step in any qualitative data analysis is importing and preparing the dataset. Traditionally, this involves manually coding large volumes of text, but with tools like ChatGPT, researchers can automate this process. For example, instead of manually setting up themes in NVivo, a researcher could interact with ChatGPT to organise and code the data.

Prompt 4.2.1.1
"I have focus group transcripts on consumer attitudes toward sustainable products. Can you help me organise and code the data based on themes like 'Environmental Concerns' and 'Product Quality'?"

In response, ChatGPT would suggest how to structure the dataset and offer advice on categorising key themes. It could also generate code to automatically classify the text using Python libraries such as NLTK or spaCy. Below is an example of how ChatGPT might assist with converting the text into a structured format for analysis:

```
import pandas as pd
from sklearn.feature_extraction.text import TfidfVectorizer

# Example dataset of focus group transcripts
data = {
    'Transcript': [
        "I think sustainable products are great for the
environment.",
        "The quality of some eco-friendly items isn't as good.",
        "I'm willing to pay more for sustainable goods, but they
need to be convenient."
    ]
}

df = pd.DataFrame(data)


# Convert text to TF-IDF features for theme-based analysis
vectorizer = TfidfVectorizer(stop_words='english')
X = vectorizer.fit_transform(df['Transcript'])
```

By generating code like this, ChatGPT helps researchers efficiently organise and code qualitative data. For predefined themes such as "Environmental Concerns" or "Product Quality," the tool can automate the coding process, while also suggesting new, emerging themes based on content analysis, allowing for a more exploratory approach to qualitative research.

4.2.2 Enhancing Analysis with AI Tools

Once the initial coding is complete, ChatGPT can assist in deepening the analysis by helping to identify patterns, conduct sentiment analysis, and even suggest additional coding schemes. Researchers who have completed the initial coding phase might prompt ChatGPT for further assistance, such as:

Prompt 4.2.2.1
"Can you help me identify common phrases and themes in the transcripts? I'm looking for any new patterns that might emerge."

In response, ChatGPT would generate code to search for recurring terms or themes across the dataset, allowing researchers to uncover insights that might not have been immediately apparent through manual coding. Additionally, ChatGPT can help perform sentiment analysis to categorise sections of the text as positive, negative, or neutral, providing further insights into participants' attitudes.

For example, ChatGPT might suggest the following Python code to conduct a text search and sentiment analysis:

```
from sklearn.feature_extraction.text import CountVectorizer
from vaderSentiment.vaderSentiment import
SentimentIntensityAnalyzer

# Text search for common phrases
vectorizer = CountVectorizer(stop_words='english')
X = vectorizer.fit_transform(df['Transcript'])
print(vectorizer.get_feature_names_out())

# Sentiment analysis
analyzer = SentimentIntensityAnalyzer()
df['Sentiment'] = df['Transcript'].apply(lambda x: analyzer.
polarity_scores(x)['compound'])
print(df[['Transcript', 'Sentiment']])
```

This AI-driven process allows researchers to identify themes such as "sustainability" or "product quality" and explore the sentiments associated with these themes. ChatGPT can also highlight emerging topics that were not predefined, offering a more exploratory approach to data analysis compared to the predefined structure of NVivo.

4.2.3 Visualising Results with AI

Visualising qualitative data is essential for effectively communicating findings. In NVivo, researchers have access to built-in visualisation tools, but with ChatGPT, researchers can generate code to create a range of custom visualisations. For instance, a researcher could ask ChatGPT:

Prompt 4.2.3.1
"Can you help me create visualisations, such as word clouds and charts, to illustrate the key themes in the focus group discussions?"

In response, ChatGPT would generate code to create visualisations like word clouds or bar charts, which summarise the frequency of key terms or themes in the dataset. Below is an example of how ChatGPT might help generate a word cloud to display the most frequently discussed topics:

```
from wordcloud import WordCloud
import matplotlib.pyplot as plt

# Generate a word cloud
text = " ".join(df['Transcript'])
wordcloud = WordCloud(stopwords='english', background_
color='white').generate(text)

# Display the word cloud
plt.figure(figsize=(8, 6))
plt.imshow(wordcloud, interpolation='bilinear')
plt.axis('off')
plt.show()
```

In addition to word clouds, ChatGPT can help create more complex visualisations, such as mind maps or sentiment distribution charts, using libraries like matplotlib and seaborn. These visualisations can summarise the data and reveal underlying patterns, providing researchers with a clearer understanding of the relationships between different themes.

By using ChatGPT to assist with qualitative data analysis, researchers can automate key steps in the process, from coding and analysing text data to generating visualisations. This approach not only saves time but also enhances the depth and accuracy of the analysis, making it easier to identify meaningful insights within large datasets. While NVivo offers a more structured environment, ChatGPT provides a flexible, customisable approach that can cater to the unique needs of each research project.

5 AI in Data Visualization

The integration of AI into data visualization has transformed how complex datasets are represented, making it easier to interpret and gain insights from the data. AI-driven tools have introduced innovations such as automated chart creation and interactive dashboards, enabling more intuitive and insightful visual representations. This section explores these innovations through examples that demonstrate how students can effectively apply AI to automate and enhance data visualization tasks.

5.1 Automated Chart Creation

Creating data visualizations, such as sales performance charts, has traditionally required manual data aggregation, selecting appropriate chart types, and formatting the visual representation, all of which can be time-consuming and prone to errors. With the integration of AI, these tasks can be automated, significantly reducing the time required while ensuring accuracy and visual appeal. Let's take a step-by-step look at how AI can automate the creation of a sales performance chart for a retail company, visualising monthly sales data across various product categories.

5.1.1 Step 1: Data Preparation

The first step in automating chart creation involves preparing the dataset. For this example, we have monthly sales figures for product categories like Electronics, Clothing, and Groceries. The dataset must be structured in a way that AI tools can easily process and visualise.

```
import pandas as pd


# Example dataset
data = {'Month': ['January', 'February', 'March',
'April', 'May'],
        'Electronics': [12000, 15000, 13000, 16000, 17000],
        'Clothing': [10000, 11000, 10500, 11500, 12000],
        'Groceries': [8000, 8500, 9000, 9500, 10000]}

df = pd.DataFrame(data)
```

In this step, the sales data is organised by month and product category, creating a structured dataset. Ensuring that the data is clean and well-organised is crucial for accurate visualisation. The integrity of the dataset must be maintained to ensure that AI tools can effectively generate accurate charts based on the data.

5.1.2 Step 2: Automating Chart Creation Using AI Tools

Once the data is prepared, AI-driven tools like Tableau, Power BI, or Python libraries such as Matplotlib can be used to automatically suggest and generate the most suitable chart types. These tools analyse the dataset and recommend visual representations that best convey the information. Here's an example using Python and Matplotlib to automatically generate a line chart for sales trends.

```
import matplotlib.pyplot as plt


# Automatically generate a line chart to show sales trends
df.plot(x='Month', y=['Electronics', 'Clothing', 'Groceries'],
kind='line')

# Display the chart
plt.title('Monthly Sales Performance')
plt.xlabel('Month')
plt.ylabel('Sales (USD)')
plt.show()
```

In this step, the AI tool analyses the data and automatically generates a line chart displaying sales trends across different product categories. This automation not only reduces the time spent manually creating the chart but also ensures that the chosen chart type is optimised for the dataset, making the visualisation clearer and more effective.

5.1.3 Step 3: Customizing and Enhancing the Chart

AI tools can also assist in customising the chart to improve its interpretability. Enhancements might include adding trend lines, annotations, or adjusting colours to highlight specific data points or trends. For example, the following code adds a trendline for Electronics sales, making the sales trajectory clearer over time.

```
# Adding a trendline for Electronics sales
df['Electronics'].plot(kind='line', label='Electronics',
color='blue', marker='o')
plt.title('Electronics Sales Trend')
```

```
plt.xlabel('Month')
plt.ylabel('Sales (USD)')
plt.grid(True)
plt.show()
```

In this step, customising the chart by adding a trendline for Electronics sales provides a clearer understanding of the data. Features such as gridlines, markers, and colour adjustments can further enhance the chart's visual appeal and clarity, making it easier to interpret the data and draw insights.

By automating the chart creation process, AI allows users to focus on interpreting the data rather than spending time on the technical aspects of chart design. This approach streamlines workflows and ensures that charts are optimised for the data they represent, making them more effective tools for communication and decision-making.

5.2 *Development of Interactive Dashboards*

Interactive dashboards have become indispensable tools in modern data visualisation, allowing users to explore and analyse data dynamically. By enabling real-time filtering, manipulation, and visualisation, these dashboards provide deeper insights and a more engaging user experience. In this section, we demonstrate how to create an interactive sales dashboard that allows users to explore sales data by filtering through product categories, time periods, and regions. AI-enhanced tools streamline the process, making it easier to develop intuitive and insightful dashboards.

5.2.1 Step 1: Data Preparation

The first step in building an interactive dashboard is to prepare the dataset. The sales data, broken down by product category, month, and region, needs to be organised to support dynamic filtering and visualisation. This structured dataset allows AI-enhanced tools to process and display the data interactively.

```
import pandas as pd


# Extended dataset for dashboard example
data = {'Month': ['January', 'February', 'March',
'April', 'May'],
        'Electronics_North': [12000, 15000, 13000, 16000, 17000],
        'Electronics_South': [13000, 14000, 13500, 15000, 16000],
        'Clothing_North': [10000, 11000, 10500, 11500, 12000],
```

```
        'Clothing_South': [9000, 10500, 11000, 11500, 11800],
        'Groceries_North': [8000, 8500, 9000, 9500, 10000],
        'Groceries_South': [8500, 8700, 9100, 9200, 9800]}

df_dashboard = pd.DataFrame(data)
```

In this step, the sales data is organised by month, product category, and region. The dataset is now ready for use in the dashboard, supporting dynamic filtering and visualisation. Ensuring the data's accuracy and completeness is crucial for generating reliable and meaningful insights through the dashboard.

5.2.2 Step 2: Using AI-Enhanced Tools to Develop the Dashboard

Once the data is prepared, AI-enhanced tools such as Tableau, Power BI, or Python's Dash library can be used to develop the interactive dashboard. These tools leverage AI features to suggest optimal visualisation types and provide a high degree of interactivity, making the dashboard more user-friendly and insightful.

Here's an example of how you might use Python's Dash library to create an interactive sales dashboard:

```
import dash
import dash_core_components as dcc
import dash_html_components as html
from dash.dependencies import Input, Output

# Initialize the Dash app
app = dash.Dash(__name__)

# Layout of the dashboard
app.layout = html.Div([
    dcc.Dropdown(
        id='category-dropdown',
        options=[{'label': 'Electronics', 'value':
'Electronics'},
                 {'label': 'Clothing', 'value': 'Clothing'},
                 {'label': 'Groceries', 'value': 'Groceries'}],
        value='Electronics'
    ),
    dcc.Graph(id='sales-graph')
])
```

```
# Callback function to update the graph based on selected
category
@app.callback(
    Output('sales-graph', 'figure'),
    [Input('category-dropdown', 'value')]
)
def update_graph(selected_category):
    filtered_data = df_dashboard[[f'{selected_category}_North',
f'{selected_category}_South']]
    figure = {
        'data': [
            {'x': df_dashboard['Month'], 'y': filtered_
data[f'{selected_category}_North'], 'type': 'line', 'name':
f'{selected_category} North'},
            {'x': df_dashboard['Month'], 'y': filtered_
data[f'{selected_category}_South'], 'type': 'line', 'name':
f'{selected_category} South'}
        ],
        'layout': {
            'title': f'{selected_category} Sales by Region'
        }
    }
    return figure
# Run the app
# app.run_server(debug=True)
```

In this step, the dashboard layout is designed with a dropdown menu that allows users to select different product categories (e.g., Electronics, Clothing, Groceries). The graph dynamically updates based on the selected category, displaying the sales trends across different regions (North and South). This type of real-time interaction is facilitated by AI-enhanced tools, which make the dashboard more engaging and easier to navigate.

5.2.3 Step 3: Enhancing Interactivity and User Experience

AI-driven tools can suggest additional interactive features to further enhance the dashboard's user experience. These features might include drill-down options, dynamic filtering, and real-time data updates, all of which contribute to a more engaging and insightful dashboard.

Several enhancements can be integrated to further improve interactivity and functionality. For example, adding a date picker allows users to filter data by specific months or time periods, offering more granular insights into sales performance. This feature enables users to focus on particular time frames and compare performance across various periods. Another useful enhancement is regional filters, which allow users to compare sales performance across different locations. By integrating filters for regions like North and South, users can identify regional trends, helping them make more informed decisions. Furthermore, implementing real-time data updates ensures that the dashboard always reflects the latest sales data, which is particularly beneficial for businesses operating in fast-paced environments.

AI-powered tools play a vital role in creating interactive dashboards that are not only visually compelling but also allow users to explore data in a dynamic, meaningful way. These dashboards serve as platforms for real-time analysis and decision-making, making complex data more accessible and actionable. By leveraging AI-enhanced features, researchers and business users alike can design dashboards that offer a rich, interactive experience, facilitating deeper insights and more informed decisions.

6 Conclusion

The integration of Artificial Intelligence into data analysis has transformed the landscape of both quantitative and qualitative research. AI technologies, particularly machine learning and deep learning, have enabled more sophisticated, efficient, and scalable methods for analyzing large and complex datasets. From automating traditional statistical methods like ANOVA and regression analysis to enhancing predictive modeling and data visualization, AI has streamlined processes that were once manual and prone to error. Furthermore, in qualitative research, AI's capabilities in clustering and topic modeling have provided new opportunities for pattern detection and thematic analysis, making sense of unstructured data in ways that were previously unimaginable. As AI continues to evolve, its applications in data analysis are poised to grow, offering deeper insights and enhancing the accuracy of decision-making across diverse fields such as healthcare, social sciences, and even natural science studies. The future of AI in data analysis holds tremendous potential not only to optimize existing methods but also to uncover new avenues for scientific discovery and innovation. Embracing these advancements will allow researchers, students, and professionals to harness AI's full potential, driving significant progress in data analysis and research methodologies.

Acknowledgements This research was supported by the Ministry of Education of Humanities and Social Science Project (No. 19YJA72040002). The author gratefully acknowledges the financial assistance provided, which made this work possible.

References

George, A. S., George, A. H., Shahul, A., & Baskar, T. (2023). AI-driven breakthroughs in healthcare: Google health's advances and the future of medical AI. *Partners Universal International Innovation Journal, 1*(3), 256–267. https://doi.org/10.5281/zenodo.8085221

Hastie, T., Tibshirani, R., & Friedman, J. (2017). *The elements of statistical learning: Data mining, inference, and prediction* (2nd ed.). Springer. https://doi.org/10.1007/978-0-387-84858-7

Patil, S., & Shankar, H. (2023). Transforming healthcare: harnessing the power of AI in the modern era. *International Journal of Multidisciplinary Sciences and Arts, 2*(1), 60–70. https://doi.org/10.47709/ijmdsa.v2i1.2513

Rahmani, A. M., Azhir, E., Ali, S., Mohammadi, M., Ahmed, O. H., Ghafour, M. Y., Hosseinzadeh, M., et al. (2021). Artificial intelligence approaches and mechanisms for big data analytics: A systematic study. *PeerJ Computer Science, 7*, e488. https://doi.org/10.7717/peerj-cs.488

Subasi, A. (2020). *Practical machine learning for data analysis using python*. Academic Press.

Chapter 6
Harnessing AI for Data Interpretation and Insightful Discussions

Lei Fang and Aniekan Essien

Abstract This chapter explores the use of AI in data interpretation and discussion, providing a step-by-step recipe with practical, hands-on instructions. We demonstrate the application of AI in interpreting both quantitative and qualitative data through two case studies. We emphasise that human researchers must lead the process, using AI as a tool that requires ongoing guidance and refinement. Although human-generated interpretations offer greater depth, link to context, and engagement with literature and theory, AI enhances the process by increasing efficiency, reducing errors, and offering fresh perspectives through human-AI collaboration.

1 Introduction

The use of artificial intelligence (AI) in academic research is changing how we gather, analyse, and understand data across various disciplines (Duan et al., 2019; Igbokwe, 2023). This chapter critically discusses how AI can support researchers by making data analysis both deeper and broader yet emphasising the importance of not letting AI work in isolation. There is a need to use AI responsibly, ensuring it complements—rather than overtakes—the human element in research (Sison et al., 2024). As these tools become more and more integrated into academic research, it is important to explore their capabilities and understand the range of responsibilities that come with their use. For example, we must consider how they are applied and the impact they have, ensuring that their integration is both thoughtful and responsible.

L. Fang (✉)
Queen Mary University of London, London, UK
e-mail: lei.fang@qmul.ac.uk

A. Essien
University of Bristol, Bristol, UK

X. Zhou, H. Al-Samarraie (eds.), *Institutional guide to using AI for research*, Advances in Artificial Intelligence in Education 2,
https://doi.org/10.1007/978-3-031-94809-1_6

The ethical use of AI in academic research necessitates a framework that promotes transparency, accountability, and fairness. Researchers must be vigilant in their use of AI, ensuring that they understand the workings of their AI tools and the provenance of the data they use. Furthermore, there should be an ongoing assessment of AI-driven research outcomes to detect and correct biases. Transparency is crucial, not merely in the research process but also in communicating the capabilities and limitations of AI to the broader academic community and the public. Integrating AI into the research process should be done thoughtfully and deliberately, with an emphasis on collaboration between AI and human intelligence. Researchers should not view AI as a replacement for traditional research methods but as an enhancement that provides additional insights and efficiencies.

This collaborative approach (e.g., Anthony et al., 2023) involves, for example:

- Defining clear roles for AI and human researchers to prevent over-reliance on AI.
- Training researchers to effectively use and understand AI tools.
- Establishing interdisciplinary teams to bridge the gap between technical AI experts and subject-matter experts.

As we advance, the role of AI in academic research will likely grow, necessitating continued vigilance and adaptation. Although AI can offer valuable insights and interpretations that we might not have considered, it is crucial that we remain in control of the research process and make informed decisions, as AI can sometimes produce unexpected or flawed results. In this chapter on data interpretation and discussion, we focus on how AI can help generate new insights, without revisiting the data analysis itself.

By the end of this chapter, students should be able to:

1. Understand the transformative role AI plays in automating data analysis, enabling researchers to handle and interpret larger datasets efficiently.
2. Recognise the challenges and limitations of AI, particularly the potential for inherent biases in AI systems and the difficulty in interpreting AI-driven conclusions.
3. Learn how to integrate AI tools responsibly into their research practices, emphasising the balance between AI capabilities and human oversight.
4. Develop both qualitative and quantitative strategies for maintaining transparency and accountability in AI-driven research to ensure ethical application and communication of findings.

2 Literature Review

AI has come a long way since its inception, evolving from simple automated tasks to complex algorithms that can learn and adapt. Today, AI is becoming an important contributor to research, helping scholars across various fields to analyse massive amounts of data quickly and with greater accuracy than ever before. This speeds up the research process as well as opens new possibilities for discoveries and insights that were previously out of reach. The use of AI in research has become essential, transforming traditional methods and allowing for more in-depth exploration and understanding of complex issues.

2.1 Key Applications of AI in Academic Research

AI is fundamentally reshaping academic research by enhancing the ability to manage and interpret vast datasets with unprecedented speed and efficiency (Dwivedi et al., 2023). This allows researchers to manage and interpret larger datasets than was previously possible, drastically speeding up the process. By processing large volumes of data swiftly and efficiently, AI helps identify patterns and trends much faster than traditional methods could. An example of this impact can be seen in genomic research, where it has significantly accelerated the analysis of extensive genetic data (Guo et al., 2023). This faster processing leads to quicker and more precise predictions about genetic disorders. It also enhances the development of personalised medicine, enabling treatments to be specifically tailored to individual genetic profiles (Vilhekar & Rawekar, 2024).

Further, AI technologies (e.g., machine learning algorithms) have revolutionised areas such as climate science, where they are used to process data from satellites and sensors to model climate changes and predict weather patterns with high accuracy (Rolnick et al., 2022). In the social sciences, AI tools are applied to analyse texts and social media data to gain insights into public opinions and societal trends, providing researchers with the ability to track changes in real-time (Gerlich et al., 2023). However, the power of AI extends beyond mere data processing. AI can uncover insights that might be too subtle or complex for human detection, thereby opening up new avenues for inquiry and exploration. This capability makes AI more than merely a tool for data processing but even a partner in hypothesis generation and theory development (e.g., Essien et al., 2024). Additionally, AI contributes to the discourse within scientific communities by suggesting alternative interpretations and fostering discussions that challenge conventional understandings (Bearman et al., 2023). This positions AI not just as a tool for data processing, but as an integral collaborator in scientific research, enhancing both the breadth and depth of investigations.

While the integration of AI into research methodologies offers vast potential, it demands a critical perspective. Researchers must embrace the efficiencies and capabilities of AI as well as rigorously examine and continually assess the methodologies and outcomes driven by AI to avoid biases and ensure the validity of their research findings. This ensures that AI serves as a powerful ally in the pursuit of knowledge, enhancing academic inquiry.

2.2 Challenges of AI in Academic Research

Despite the rapid integration of artificial intelligence (AI) into research methodologies across various fields, there are significant challenges that need critical attention to ensure ethical and reliable outcomes. The inherent biases in AI systems (Ferrara, 2024) are a primary concern. These biases originate from the training datasets, which may contain historical prejudices that AI can unknowingly learn and perpetuate (Ntoutsi et al., 2020). The replication of such biases can lead to skewed results and potentially harmful decisions, especially in sensitive areas like criminal justice and healthcare where they could affect individual lives and societal norms significantly (Giovanola & Tiribelli, 2023).

The issue of bias in AI also extends to the socioeconomic impacts, where algorithms could inadvertently widen existing inequalities. This is seen in job screening processes where AI might favor certain demographics over others based on biased historical hiring data (Fabris et al., 2023). Thus, bias is as much a technical challenge as a moral imperative that requires ongoing scrutiny and corrective measures.

Another pressing issue is the opacity of many AI models, commonly referred to as the "black box" problem (Brożek et al., 2024). This term describes how some AI systems operate in ways that are not transparent, making it difficult for researchers to understand how decisions are made. This lack of transparency can hinder the validation and replication of research findings, fundamental aspects of scientific inquiry (Durán & Jongsma, 2021). Transparency is crucial for maintaining trust in AI-enhanced research, as it enables peer reviewers and the broader academic community to verify and critique the workings and outputs of AI systems (Ekundayo et al., 2024). Without transparency, the scientific community may struggle to fully embrace AI tools, no matter how sophisticated they are (Vössing et al., 2022).

Researchers are therefore tasked with developing methodologies that leverage the capabilities of AI technologies, as well as also address the ethical and operational challenges listed. This involves creating AI systems that are built with ethical considerations from the ground up, ensuring fairness and transparency (Díaz-Rodríguez et al., 2023). Moreover, the academic community must foster an environment where ethical AI use is a fundamental practice, encouraging open discussions and collaborative efforts to tackle these challenges. To conclude, while AI presents a remarkable tool for advancing research, its application must be approached with caution, balancing innovation with a firm commitment to ethical practices and transparency. Researchers must remain vigilant and proactive in addressing the ethical challenges posed by AI, ensuring that their use of such technology reflects both scientific and societal values.

3 Recipe for Using AI in Data Interpretation and Discussion

In this section, we propose a recipe outlining the steps to effectively use AI for interpreting and discussing data.

Step 1. Define research questions and preparation: We begin by clearly defining our research questions and objectives. This clarity is crucial as it guides how AI can be effectively employed to address specific areas of interest. We prepare our data for AI analysis by ensuring it is correctly formatted and annotated, if necessary. This might involve structuring data into a format that AI tools can process.

Step 2. Implement AI for data interpretation: For quantitative data, AI can help to identify subtle patterns or trends, which can be especially useful in large datasets where manual inspection might miss important details. AI can also be used to identify outliers or anomalies, which could indicate significant but rare events or errors in data collection. Additionally, based on existing data trends, AI can offer predictive insights, suggesting potential future trends.

For qualitative data, we use AI tools to perform initial data coding and to identify themes. AI can quickly sift through large volumes of text, detecting patterns and recurring themes that might not be immediately evident. By applying natural language processing (NLP) techniques to parse and understand the nuances within the data, AI helps us further refine the thematic categories.

In this step, human researchers *lead* the process, with AI serving an assistant role in interpretation. This is reflected in several key aspects: (1) The AI-assisted data interpretation process is ***iterative***, requiring researchers to coach the AI, correcting its errors and refining its performance. (2) It is essential to provide AI with ***contextual information***, such as relevant theories, to ensure it accurately discusses the data analysis results within the proper context. (3) AI may occasionally offer ***novel or unexpected insights***, complementing and enriching existing interpretations. However, it is the human researcher who decides how to incorporate these insights within the discussion.

Step 3. Synthesise AI and human insights: We integrate AI-derived insights with human analysis to formulate a coherent discussion. While AI excels at identifying broad patterns, human researchers are necessary to interpret these patterns within the correct cultural and contextual frameworks. We should conduct a thorough review of the AI-generated interpretation alongside traditional methods to ensure the validity and reliability of the findings.

Step 4. Evaluate AI's role, and address ethical and transparency issues: Finally, we critically evaluate the contributions of AI to the research process. We assess areas where AI added value and identify any aspects it may have overlooked or misinterpreted. We refine the AI analysis based on initial findings and feedback from the iterative review, adjusting AI parameters or redefining algorithms to align closely with specific research needs. We should ensure transparency in the AI processes used, and document how AI tools were implemented and any decisions made based on AI insights. We discuss potential biases introduced by AI and the steps taken to mitigate them. Throughout, we uphold ethical standards, particularly regarding data privacy and the integrity of the analysis.

Case Study 1: Using AI for Quantitative Data Interpretation
Applying the recipe suggested above, we explore a quantitative case study, followed by a qualitative one, to demonstrate the application of AI in both types of data interpretation. First, we explore how AI can assist in interpreting numerical data while ensuring that researchers retain a critical and active role throughout the process.

Research context and objectives: This case study interprets the data analysis results presented in Essien et al. (2024), which examines the impact of AI text generators on critical thinking skills across various levels of Bloom's taxonomy (Bloom, 1956; Forehand, 2010). The study hypothesises that AI text generators could enhance critical thinking by offering diverse perspectives and promoting deeper cognitive engagement. Using pre-test and post-test data from a controlled experiment, the study seeks to validate this hypothesis. Here, we focus exclusively on interpreting the results presented in Table 6.1 (the original Table 4 from Essien et al., 2024).

Step 1. Define research questions and preparation

According to Essien et al. (2024), the research question concerning the results in Table 6.1 is: What is the impact of AI text generators on critical thinking skills as measured by Bloom's taxonomy? The hypothesis is that the use of AI text generators leads to statistically significant improvements in critical thinking skills.

Step 2. Implement AI for data interpretation.

We utilise ChatGPT-4 to assist our data interpretation. We provide ChatGPT with the specific research question and hypothesis we aim to test, which are outlined in Hypothesis Statement 1 from Essien et al. (2024). We also inform ChatGPT that we

Table 6.1 Paired-Sample t-test results for intervention impact on Bloom's taxonomy levels

Bloom's taxonomy level	Test pair	Mean before intervention	Mean after intervention	t-value	df	p-value
Remember	Q9–Q23	3.15	4.07	−7.36	105	<0.001*
	Q10–Q24	4.12	4.96	−6	105	<0.001*
Understand	Q11–Q25	3.92	4.69	−9.25	105	<0.001*
	Q12–Q26	4.15	4.54	−5	105	<0.001*
Apply	Q13–Q27	4.07	4.44	−6.52	105	<0.001*
	Q14–Q28	4.15	4.82	−7.2	105	<0.001*
Analyse	Q15–Q29	4.15	4.42	−5.12	105	<0.001*
	Q16–Q30	3.92	4.34	−5	105	<0.001*
Evaluate	Q17–Q31	3.83	4.24	−5.32	105	<0.001*
	Q18–Q32	2.96	3.28	−4.4	105	<0.001*
Create	Q19–Q33	2.98	3.22	−1.77	105	0.079
	Q20–Q34	3.13	3.36	−2	105	0.08

Notes: *denotes $p < 0.05$, indicating a statistically significant difference between the control and intervention groups for that question. This is the data analysis results to be used for AI-assisted interpretation, from Table 4 of Essien et al. (2024)

are using Bloom's Taxonomy as the theoretical framework. We upload Table 6.1 into the ChatGPT interface with the following instructions in the prompt:

> Please interpret the result from the uploaded table to address the following research question: What is the impact of AI text generators on critical thinking skills as measured by Bloom's taxonomy? The hypothesis is that the use of AI text generators leads to statistically significant improvements in critical thinking skills. Bloom's Taxonomy is the theoretical framework for this analysis.

Table 6.2 shows the interpretation generated by ChatGPT, together with the human-generated interpretation by the researchers of the paper. When comparing the interpretations provided by ChatGPT with those crafted by human researchers, several observations can be made. Both AI and human interpretations provide consistent insights into key metrics like mean scores, t-values, and statistical significance. However, human interpretation offers deeper insights than ChatGPT. (1) Human interpretation groups Bloom's taxonomy levels (e.g., lower level: Remember, Understand; higher level: Apply, Create) into broader categories, offering a more synthesised view. (2) The researchers link the underlying theory for the interpretation, which ChatGPT lacks. (3) The researchers also connect the results to other contextual information, e.g., Table 3 in Essien et al. (2024). (4) ChatGPT's interpretation focuses primarily on statistical significance without exploring how substantial the changes might be. In contrast, the researchers delve into the nuances of effect sizes, highlighting the varied impacts of the intervention on different cognitive skills.

Now, let's iterate the process and coach ChatGPT to refine its interpretation. We instruct it to categorise the results into lower and higher levels based on Bloom's Taxonomy. Additionally, we provide Table 3 from Essien et al. (2024) for a more comprehensive context. We put the following instructions in the prompt:

> Thank you. Please categorise the results into lower and higher levels according to Bloom's Taxonomy. This new table just uploaded provides additional information for your reference.

Table 6.3 presented below shows the enhanced interpretation produced by ChatGPT. ChatGPT, following instructions, refined its method by categorising results into lower and higher cognitive levels. It specifically highlights substantial improvements in the foundational cognitive skills of Remembering and Understanding, aligning well with the human interpretation of the intervention's effectiveness in enhancing these basic cognitive abilities. Both interpretations agree on the intervention's overall effectiveness across various cognitive skills. However, in comparison to its initial attempt, the revisions in ChatGPT's interpretation show

Table 6.2 A comparison of the interpretation generated by ChatGPT and the human-generated interpretation

Interpretation generated by ChatGPT	Human interpretation by Essien et al. (2024)
The table presents the results of a paired-sample t-test to evaluate the impact of an intervention on different levels of Bloom's taxonomy. The hypotheses tested are as follows: H0: No significant difference exists between pre-test and post-test mean scores in the intervention group. H1: A statistically significant difference is observed. Key Results Interpretation: • **Remember** (Q9–Q23 and Q10–Q24): Q9–Q23: The mean score before the intervention was 3.15, and after the intervention, it increased to 4.07. The t-value is −7.36 with a p-value of less than 0.001, indicating a statistically significant improvement. Q10–Q24: The change from a mean of 4.92 to 4.96 is not statistically significant (t-value = −1.00, p-value = 0.320), meaning no meaningful improvement occurred here. • **Understand** (Q11–Q25 and Q12–Q26): Q11–Q25: Mean before was 3.89 and after was 4.07. The t-value is −3.17, and the p-value is less than 0.001, indicating a significant improvement in understanding. Q12–Q26: The mean score also showed significant improvement from 4.07 to 4.54 ($t = -3.81$, $p < 0.001$). • **Apply** (Q13–Q27): The mean score improved from 4.16 to 4.54, with a t-value of −2.72 and a p-value of 0.005, indicating a statistically significant improvement. • **Analyze** (Q15–Q28 and Q16–Q29): Q15–Q28: The mean score rose from 3.17 to 3.78, with a highly significant t-value of −5.02 and $p < 0.001$. Q16–Q29: The increase from 3.33 to 3.77 also shows statistical significance ($t = -3.81$, $p < 0.001$). • **Evaluate** (Q17–Q30 and Q18–Q31): Q17–Q30: The mean score changed from 4.14 to 4.47, with a significant t-value of −3.37 ($p = 0.001$). Q18–Q31: The increase from 4.29 to 4.54 is also significant ($t = -3.91$, $p < 0.001$). • **Create** (Q19–Q32 and Q20–Q33): Q19–Q32: The improvement from 2.98 to 3.33 is statistically significant ($t = -2.40$, $p = 0.018$). Q20–Q33: The increase from 2.98 to 3.22 was not statistically significant ($t = -1.77$, $p = 0.079$).	Table 6.4 presents paired-sample t-test outcomes assessing intervention impact on participants' cognitive performance across Bloom's Taxonomy. The table displays key statistics for each test pair, which consists of a pre-intervention question (e.g. Q9) and a corresponding post-intervention question (e.g. Q23), as well as the mean scores before and after the intervention, t-values, degrees of freedom (df), and p-values. To illuminate the implications of these statistical findings, we provided an analysis of specific test pairs from Table 6.3. The intervention significantly increased participants' mean scores on *Remember-level* questions (3.15 to 4.07) with a highly significant difference ($t = -7.36$, $p < 0.001$), showcasing its profound impact on factual recall. *Understand-level* questions displayed noteworthy improvement (3.19–4.07), a highly significant difference ($t = -6.76$, $p < 0.001$), indicating the intervention's pivotal role in enhancing comprehension abilities. *Apply-level* questions saw moderate improvement (4.15 to 4.42), with a statistically significant difference ($t = -2.72$, $p = 0.008$*). Notably, the effect size is smaller compared to remember and understand levels, suggesting a more modest impact on application. *Create-level* questions showed a significant increase (2.98 to 3.28) with a smaller effect size ($t = -2.40$, $p = 0.018$*), implying a relatively milder impact on creative thinking skills. The findings, presented in Table 6.3, collectively underscore the multifaceted impact of the intervention across the different levels of cognitive learning. The results highlight substantial improvements in participants' abilities to remember facts, comprehend concepts, and apply knowledge. However, the extent of these improvements varies across Bloom's taxonomy levels, with a more pronounced impact observed at the lower-order cognitive levels (remember and understand) compared to the higher-order levels (apply, analyse, evaluate, and create). This insight is crucial for educators and curriculum developers tailoring strategies to specific cognitive domains.

Table 6.3 The enhanced interpretation produced by ChatGPT

Updated interpretation generated by ChatGPT after being coached
Updated Interpretation of the Results This table provides additional information on the pre-test and post-test mean scores along with their standard deviations for each test item, which helps refine the analysis of the intervention's effect on different levels of Bloom's taxonomy. Below is the updated synthesis, incorporating these values.
Lower Levels (Remember and Understand) **Remember:** **Q9 (Rem):** Pre-test Mean: 3.15 (SD = 1.21), Post-test Mean: 4.76 (SD = 0.06) There is a substantial improvement in the mean score, indicating that participants significantly improved their ability to recall information after the intervention. The low post-test SD indicates consistent improvement across participants. **Q10 (Rem):** Pre-test Mean: 4.92 (SD = 0.55), Post-test Mean: 5.00 (SD = 0.00) Although the post-test score is near perfect, the lack of a significant t-test result in Table 6.4 suggests that this improvement is not statistically significant. The small difference in pre- and post-test means and the perfect post-test score indicate that participants may have already mastered this item pre-intervention. Understand: **Q11 (Und):** Pre-test Mean: 3.19 (SD = 1.45), Post-test Mean: 4.80 (SD = 0.06) There is a large gain in understanding (mean increase of 1.61 points), with low variability in post-test scores. The statistically significant result in Table 6.4 and this improvement in mean indicates the intervention was effective in enhancing understanding. **Q12 (Und):** Pre-test Mean: 4.07 (SD = 1.27), Post-test Mean: 4.72 (SD = 0.13) The moderate increase in mean score with a reduction in variability suggests that the intervention was effective, though not as dramatically as for Q11.

(continued)

Table 6.3 (continued)

Updated interpretation generated by ChatGPT after being coached
Higher Levels (Apply, Analyze, Evaluate, and Create) **Apply:** **Q13 (App):** Pre-test Mean: 4.15 (SD = 1.18), Post-test Mean: 4.58 (SD = 0.09) The moderate increase in scores, coupled with a statistically significant t-test result, suggests a meaningful improvement in applying knowledge, though the change is smaller compared to lower levels. **Q15 (App):** Pre-test Mean: 3.18 (SD = 1.36), Post-test Mean: 4.52 (SD = 0.13) A larger increase in mean score here compared to Q13 indicates that participants improved their ability to apply concepts more effectively after the intervention, supported by the significant t-test result. Analyze: **Q16 (Ana):** Pre-test Mean: 2.93 (SD = 1.05), Post-test Mean: 4.06 (SD = 0.13) A significant improvement in analysis skills is evident, with a mean increase of over 1 point, suggesting that the intervention successfully enhanced participants' analytical abilities. **Q17 (Ana):** Pre-test Mean: 3.14 (SD = 1.17), Post-test Mean: 3.94 (SD = 0.09) A smaller improvement than Q16, but still statistically significant. The smaller standard deviation post-intervention suggests greater consistency in analytical performance after the intervention. Evaluate: **Q18 (Eva):** Pre-test Mean: 3.14 (SD = 1.33), Post-test Mean: 3.90 (SD = 0.11) A moderate improvement in evaluation skills is observed, supported by the significant t-test result. The post-test SD indicates consistent performance gains among participants. **Q19 (Eva):** Pre-test Mean: 3.02 (SD = 1.50), Post-test Mean: 3.78 (SD = 0.16) Similar to Q18, this test pair also shows significant improvement, although the larger pre-test SD suggests that participants began with varying levels of ability. Create: **Q20 (Cre):** Pre-test Mean: 2.98 (SD = 1.30), Post-test Mean: 3.72 (SD = 0.20) A significant improvement in creative skills, with the post-test mean showing a substantial increase. The high pre-test SD indicates initial variability in creativity, which was reduced after the intervention. **Q21 (Cre):** Pre-test Mean: 2.98 (SD = 1.42), Post-test Mean: 3.64 (SD = 0.21) While there is some improvement, the t-test result in Table 6.4 shows this was not statistically significant, indicating that the intervention may have had a less consistent effect on this specific aspect of creativity.

(continued)

Table 6.3 (continued)

Updated interpretation generated by ChatGPT after being coached
Overall Summaries: **Lower Levels (Remember and Understand):** There were significant improvements in almost all test pairs for the lower levels, especially in "Understand" (Q11 and Q12). The consistent post-test results across participants, as reflected by lower post-test SDs, indicate that the intervention was highly effective at improving memory and understanding. **Higher Levels (Apply, Analyze, Evaluate, Create):** The intervention also led to significant gains across all higher-order cognitive skills, particularly in applying, analyzing, and evaluating knowledge. However, creative abilities showed more variability, with only Q20 showing a statistically significant improvement. The large pre-test SDs for the higher levels suggest more variability in participants' initial skills, which was somewhat reduced post-intervention, indicating that the intervention helped bring participants to a more uniform level of competence. In summary, the intervention was highly effective in improving both lower- and higher-order cognitive skills, though certain aspects of creativity were less impacted. The low post-test standard deviations across the board indicate that the intervention generally led to consistent performance improvements among participants.

only marginal improvements in these aspects. ChatGPT provides a notably more detailed, yet somewhat limited interpretation focused strictly on statistical results for each test item. In contrast, the interpretation by human researchers links the statistical outcomes to educational strategies and broader implications.

Step 3. Synthesise AI and human insights.

In this step, we guide ChatGPT to gain further insights derived from its initial data interpretation. We provide the interpretation generated in Step 2. For robustness, we upload both the human-generated and ChatGPT-generated versions, and the insights produced by ChatGPT are largely similar across both. We also ensure that these further insights remain grounded in Bloom's Taxonomy and are informed by the full context, including the research question and hypothesis. We input the following instructions in the prompt:

> Based on the uploaded data interpretation, please conduct a discussion and provide further insights. They remain grounded in Bloom's Taxonomy. The original research question, hypothesis, and the two uploaded tables remain unchanged and serve as the context.

In Table 6.4, we present the ChatGPT-generated insights. We also incorporate relevant human discussions. In comparing the insights and discussions generated by ChatGPT with those by human researchers, we find that both recognise the impact of interventions on cognitive skills, particularly at foundational levels, emphasising their significance in educational settings. However, the analyses differ in nature and depth. (1) ChatGPT offers a detailed examination of cognitive skill improvements across Bloom's taxonomy, identifying inconsistencies and areas needing intervention, which are less explicit in the human analysis. In contrast, human researchers

Table 6.4 A comparison of ChatGPT-generated insights and human-generated insights

Insights and discussion generated by ChatGPT	Insights and discussion generated by human researchers in Essien et al. (2024)
Several key insights emerge for discussion: 1. **Significant Improvement in Lower-Level Cognitive Skills**: The intervention shows a clear positive effect on foundational cognitive skills, particularly in the areas of **Remember** and **Understand**. Test items such as Q9, Q10, and Q11 exhibit substantial improvements in post-test scores, with low variability among participants, indicating consistent learning outcomes. This aligns well with educational practices that emphasize the importance of mastering basic recall and comprehension as a foundation for higher-level thinking. 2. **Statistical Validation of Results**: The statistically significant differences in several test items, particularly at the lower levels, support the effectiveness of the intervention. For instance, Q11 (Understand) and Q12 (Understand) both show marked gains with low variability, suggesting that the intervention was particularly effective in improving comprehension skills. 3. **Moderate Gains in Higher-Order Cognitive Skills**: At the higher levels of Bloom's taxonomy—**Apply**, **Analyze**, and **Evaluate**—participants showed significant improvements. Test items like Q16 (Analyze) and Q15 (Apply) demonstrated notable gains, but the extent of improvement was smaller compared to the lower levels. This suggests that while the intervention was effective, more intensive or targeted strategies might be needed to deepen higher-order cognitive abilities. 4. **Mixed Results in Creativity**: The most variability was observed in the **Create** category, where Q20 showed a significant improvement, but Q21 did not. The higher pre-test standard deviations in these areas indicate that participants started with varying levels of creative ability, and while the intervention helped in reducing this variability, the impact was less pronounced compared to other cognitive skills. This could indicate a need for further refinement in how creativity is fostered during interventions. 5. **Consistency Across Participants**: The lower post-test standard deviations across the board highlight the intervention's ability to produce consistent improvements among participants. This suggests that the program or instructional method used had a uniform effect, benefiting a wide range of learners and raising overall competency levels. 6. **Educational Implications**: The findings suggest that the intervention is highly effective for improving foundational knowledge and understanding, which are essential for deeper cognitive processing. However, in terms of applying and creating knowledge, further instructional support may be necessary to ensure that learners fully develop these skills. This insight is crucial for educators aiming to tailor interventions for comprehensive cognitive development. In conclusion, while the intervention was highly effective in improving both lower- and higher-order cognitive skills, there is room for improvement in fostering creativity. The statistically validated improvements across various cognitive skills suggest that such interventions can significantly enhance learning outcomes, though targeted adjustments might be required to address specific areas such as creativity.	**5.1.1. Basic level (Remember, understand, apply)** The findings from our empirical data suggested that there are significant improvements in the basic levels of critical thinking using AI text generators, such as the ability to recall information (Remember) and understand complex concepts (Understand). Although the 'Apply' category also demonstrated progress, the study highlighted that the main use was in the practical application and task efficiency rather than critical thinking per se. At the basic cognitive level, participants found ChatGPT useful for streamlining research processes and mundane tasks like proofreading and structuring academic assignments. While these activities primarily fall under task efficiency, they indirectly support critical thinking by freeing up time and cognitive resources that can be allocated to more complex reasoning tasks. **5.1.2. Advanced level (analyse, evaluate, create)** The analysis showed moderate (yet significant) improvements in advanced aspects of critical thinking, although no statistically significant impact was found on the highest level (i.e., 'Create' level) concerning the development of entirely new and original work. A majority of participants reported that ChatGPT helped them in the 'analyse' and 'evaluate' components of critical thinking. Specifically, ChatGPT exposed them to new perspectives, helped in brainstorming new ideas, and even instigated a heightened awareness of the necessity for fact-checking due to the potential for 'AI hallucination.' The cross-level interactions in this study provide an intriguing illustration of how technological tools, such as ChatGPT, can function not just as a utility but as a comprehensive facilitator of cognitive growth (QAA 2023). Traditionally, the cognitive activities described in the Bloom's Taxonomy are treated as hierarchical stages, each building on the one before it. However, the cross-level interactions suggest a more dynamic, symbiotic relationship between the basic and advanced levels of critical thinking when mediated by technology. For example, lower-level activities like proofreading, summarising, and structuring often act as prerequisites for more advanced intellectual endeavours such as analysing, evaluating, and creating. These are not isolated tasks but interdependent steps in the learning and research process. What the study highlights is that GAI, via a tool such as ChatGPT, can streamline this interconnected cognitive landscape. In the fast-paced academic environment, time is a scarce resource. By saving time on basic tasks, GAI tools can allow students the luxury of time, which can be the deciding factor in whether or not higher-order thinking even occurs. Therefore, ChatGPT essentially serves as a cognitive 'force multiplier.' It doesn't merely add to the students' capabilities at different levels but multiplies their efficiency and effectiveness across the board. This has profound implications for educational practices, potentially reshaping how we think about the role of AI in learning and critical thinking development.

discuss how AI text generators enhance critical thinking. They suggest that AI could serve as a "force multiplier" in education, and thus influencing higher-level cognitive processes as well. (2) Human researchers also connect their findings to previous studies, providing a comprehensive view of AI's role in education, a context ChatGPT does not fully explore. (3) ChatGPT, however, highlights mixed results in creativity enhancement with specific examples, adding detail not covered by the human discussion.

To address the identified limitations in the discussion generated by ChatGPT, we provide additional guidance, instructing it to incorporate references to existing research and increase the depth of its analysis. We input the following instruction:

> The current discussion lacks depth. Please incorporate references to existing research and conduct a more thorough discussion.

Table 6.5 shows the enhanced discussion produced by ChatGPT following this round of coaching. While ChatGPT has updated its discussion accordingly, several persistent issues compromise the quality of the discussion. (1) The references used are not only outdated but also poorly aligned with the focus of the research on AI text generators—a field that has only recently gained significant attention. (2) ChatGPT continues to overly rely on statistical data without extracting or discussing deeper insights. (3) Its discussion lacks clarity and precision, making it too vague. Despite multiple iterations and targeted feedback aiming to refine ChatGPT's contributions, these fundamental issues remain unaddressed.

However, there are some interesting insights generated by ChatGPT that are not covered by human discussion, such as the suggestion that "***learners may require more scaffolding or peer collaboration to fully develop these higher-order skills… Instructors could integrate more collaborative and inquiry-based approaches to better support cognitive development.***" These insights, while potentially seen as irrelevant, could offer novel perspectives that have not yet been fully considered by human researchers. For example, collaboration among learners, teachers, peers, and AI could be an effective approach to enhancing critical thinking skills by encouraging learners to reflect on and critique the thought produced by their collaborators.

Step 4. Evaluate AI's role and address ethical and transparency issues.

Table 6.6 summarised a comparison between the interpretations provided by ChatGPT and those generated by human researchers. Let's reflect on the role AI has played in this case. AI, represented here by ChatGPT, serves as an ***analytical assistant***, adeptly summarising basic statistical facts. It effectively manages the quantitative burdens of research, freeing human researchers to delve into areas requiring deeper insight and nuanced understanding. AI also acts as an ***idea generator***, bringing to light insights previously unconsidered by human researchers. Additionally, it functions as a ***scrutiniser***, offering a consistent level of analysis and reducing the influence of human biases that might otherwise skew the interpretation of data.

Table 6.5 Enhanced discussion generated by ChatGPT

Updated discussion generated by ChatGPT after being coached
1. Significant Improvement in Lower-Level Cognitive Skills: The intervention demonstrated a notable impact on foundational cognitive skills, particularly in the domains of Remember and Understand. Test items such as Q9 (Remember) and Q11 (Understand) exhibited substantial improvements in post-test scores, with low variability across participants. This consistency suggests that the intervention effectively facilitated recall and comprehension, which aligns with the well-established theory of scaffolded learning (Wood et al., 1976). According to cognitive load theory (Sweller, 1988), mastery of these basic cognitive skills reduces the extraneous load on working memory, enabling learners to engage with more complex material more efficiently. In particular, low post-test standard deviations underscore the uniform success of the intervention in enhancing these lower-order skills, reinforcing findings from similar educational interventions that emphasize recall and comprehension as foundational skills (Bloom, 1956). **2. Statistical Validation of Results:** The statistically significant differences observed in the lower cognitive domains further validate the intervention's efficacy. Items like Q11 (Understand) and Q12 (Understand) show marked gains with minimal variability. These outcomes are consistent with previous research that highlights the strong relationship between guided instruction and improved comprehension skills (Hattie, 2009). The significant t-values and low p-values provide empirical support for the intervention's success, affirming its robustness in a controlled learning environment. According to Bloom's (1956) taxonomy, comprehension is an essential precursor to higher-order thinking, and these improvements suggest that learners are well-prepared to advance to more complex cognitive tasks. **3. Moderate Gains in Higher-Order Cognitive Skills:** While there was significant improvement in higher-order skills—Apply, Analyze, and Evaluate—the gains were more moderate compared to the foundational levels. Test items like Q15 (Apply) and Q16 (Analyze) demonstrated notable but smaller increases, suggesting that while the intervention positively impacted these skills, deeper engagement is needed to maximize outcomes. Research in instructional design, particularly Merrill's (2002) first principles of instruction, emphasizes the need for problem-centered learning environments to effectively foster higher-order thinking. These findings suggest that while traditional interventions may enhance understanding and application, more complex instructional methods such as project-based learning or inquiry-based approaches may be needed to foster deeper cognitive engagement (Barron & Darling-Hammond, 2008). **4. Mixed Results in Creativity:** The intervention's impact on creative skills, as seen in the Create domain, was mixed. Q20 showed significant improvement, but Q21 did not. This variability could be explained by the diverse baseline levels of creativity among participants, as evidenced by higher pre-test standard deviations. Creativity, as conceptualized by Torrance (1974), often requires not only structured learning but also opportunities for exploration and autonomy. The findings suggest that while the intervention reduced variability in creative abilities, further enhancements may require strategies that specifically nurture creative thinking. Research indicates that fostering creativity in educational settings benefits from less structured, more student-driven approaches (Amabile, 1996), which may not have been fully integrated into this intervention. **5. Consistency Across Participants:** The low post-test standard deviations across most test items suggest that the intervention successfully produced consistent improvements among participants. Such uniformity is important, as it indicates that learners from diverse backgrounds or with varying pre-existing knowledge levels benefited from the program. This aligns with the principles of universal design for learning (UDL), which posits that instructional interventions should be flexible enough to accommodate the diverse learning needs of all students (Rose & Meyer, 2002). The consistency observed suggests that the intervention adhered to UDL principles, making it adaptable and effective for a broad audience. **6. Educational Implications:** The findings of this study have significant implications for educational practice. The clear improvements in foundational knowledge, especially in the Remember and Understand categories, highlight the intervention's efficacy in enhancing essential cognitive skills. However, in higher-order skills like Apply and Create, the intervention shows potential but could benefit from more targeted strategies to deepen cognitive engagement. Vygotsky's (1978) zone of proximal development suggests that learners may require more scaffolding or peer collaboration to fully develop these higher-order skills. Instructors could integrate more collaborative and inquiry-based approaches to better support cognitive development in these areas, in line with constructivist theories of learning (Bruner, 1966). **Conclusion:** The intervention demonstrated a strong impact on both lower- and higher-order cognitive skills, with statistically validated improvements across multiple cognitive domains. While foundational skills like recall and comprehension showed consistent and significant improvement, higher-order skills like creativity exhibited more variability. This suggests that while the intervention was highly effective, further refinement is needed to optimize outcomes in specific areas like creativity. Drawing from existing educational research, targeted adjustments such as incorporating inquiry-based learning or problem-centered approaches could help bolster the development of these more complex cognitive skills. **References:** • Amabile, T. M. (1996). *Creativity in Context.* Westview Press. • Barron, B., & Darling-Hammond, L. (2008). Teaching for meaningful learning: A review of research on inquiry-based and cooperative learning. *Powerful Learning: What We Know About Teaching for Understanding.* • Bloom, B. S. (1956). *Taxonomy of Educational Objectives: The Classification of Educational Goals.* Longmans. • Bruner, J. S. (1966). *Toward a Theory of Instruction.* Harvard University Press. • Hattie, J. (2009). *Visible Learning: A Synthesis of Over 800 Meta-Analyses Relating to Achievement.* Routledge. • Merrill, M. D. (2002). First principles of instruction. *Educational Technology Research and Development*, 50(3), 43–59. • Rose, D. H., & Meyer, A. (2002). *Teaching Every Student in the Digital Age: Universal Design for Learning.* ASCD. • Sweller, J. (1988). Cognitive load during problem solving: Effects on learning. *Cognitive Science*, 12(2), 257–285. • Torrance, E. P. (1974). *The Torrance Tests of Creative Thinking Norms-Technical Manual.* Scholastic Testing Service. • Vygotsky, L. S. (1978). *Mind in Society: The Development of Higher Psychological Processes.* Harvard University Press. • Wood, D., Bruner, J. S., & Ross, G. (1976). The role of tutoring in problem-solving. *Journal of Child Psychology and Psychiatry*, 17(2), 89–100.

Table 6.6 A comparative analysis between the interpretations provided by ChatGPT and those generated by human researchers like Essien et al. (2024)

Aspect	ChatGPT interpretation	Human interpretation
Focus	Statistical analysis and direct interpretations of cognitive skill improvements.	Broader contextual analysis linking AI'S impact to educational practices and cognitive theory.
Depth of insight	Tends to focus on direct results from data; less synthesis of broader implications.	Integrates findings into broader educational and cognitive contexts.
Citation and link to other research	Limited use of updated and relevant references; occasionally outdated or irrelevant citations.	Utilises current research effectively to contextualize findings.
Discussion clarity	Can be vague and overly focused on statistical outcomes without clear actionable insights.	Provides clear, actionable insights that are linked to practical educational strategies.
Innovativeness	Offers unique perspectives occasionally but struggles with integrating these into a cohesive discussion.	Presents innovative ideas on integrating AI into learning processes and cognitive development.

Case Study 2: Using AI for Qualitative Data Interpretation

In the preceding discussions, we have seen how AI can enhance quantitative research by rapidly processing and analysing numerical data, enabling deeper insights and interpretations. Now, we turn our focus to qualitative research, examining how AI can assist in uncovering themes, patterns, and deeper meanings within non-numerical data. This section will outline a structured approach for integrating AI into qualitative research, followed by a case study that illustrates this method in practice.

- ChatGPT-assisted inductive thematic analysis

  Introduction to the case: This case study explores a methodological innovation where GenAI, specifically ChatGPT, is integrated with traditional qualitative research techniques to perform inductive thematic analysis (Perkins & Roe, 2024). The aim of the study was to demonstrate how AI can assist in deepening qualitative analysis while maintaining the rigorous standards necessary for academic research.

  Research context and objectives: The study was aimed at identifying the potential roles that GenAI tools can play in enhancing the process of thematic analysis within qualitative research. Specifically, it examined how these tools could work in tandem with human researchers to analyse educational policies regarding the use of GenAI in research publications.

  Methodology: The study methodology involved two researchers independently analysing a dataset: one using traditional manual techniques and the other using the premium version of ChatGPT (GPT-4) to assist in the analysis. This dual approach allowed for a direct comparison of insights derived from both methods and provided a comprehensive understanding of the thematic elements within the data.

AI implementation and data analysis: ChatGPT was employed to initially code and categorise the data, identifying themes based on the frequency and context of terms within the dataset. This AI-driven process was noted for its efficiency in handling large volumes of data, which enabled a rapid preliminary analysis that highlighted potential areas for deeper exploration by the human researcher (see Table 6.7).

Challenges and validation: The key challenges encountered included managing inconsistencies in AI outputs, which sometimes presented as 'hallucinations'—data or conclusions not supported by the actual dataset. This required the human researcher to perform rigorous validation checks to ensure the integrity of the research findings. Moreover, the 'black box' nature of the tool made it difficult to trace how certain conclusions were derived, underscoring the need for a transparent and interpretable AI use in academic research.

Integration of AI and human insights: After the initial AI-assisted analysis, the insights were critically reviewed by the human researcher to interpret and contextualise them within the broader scope of the research objectives. This step was crucial in maintaining the depth and quality of the thematic analysis, ensuring that the AI's computational power was effectively harnessed to support, not replace, human expertise.

Ethical considerations/transparency: The study also addressed the ethical implications of using AI in research, particularly the risks of bias and the

Table 6.7 Comparison of traditional versus ChatGPT supported analysis, from Table 1 of Perkins and Roe (2024)

Step	Traditional analysis	ChatGPT-assisted analysis
Data familiarization	Collection of data and re-reading of data, noting down initial ideas in word processing software.	
Generating initial codes	Systematic coding of data using traditional software (MaxQDA).	Writing and re-writing prompts to instruct the GenAI tool what is required from the output.
Searching for themes	Researcher collates codes into prospective themes.	Inspecting the themes given by the GenAI tools and cross-referencing with examples from the dataset for validation.
Reviewing themes	Inspecting the themes and reviewing their relation to the codes and dataset.	Re-running the analysis at different points in time to enhance the validity of themes.
Defining and naming themes	Continuing analysis and specification of each theme.	Enhancing theme validity by repeating the analysis at multiple points.
Combining and contrasting results	Results compared by both researchers side by side, congruence identified in themes and datasets, final set of themes produced.	Themes and datasets reviewed to finalize a set of themes, focusing on consistency and relevance.
Producing the report	Creating the scholarly output of the paper.	

potential impact on research integrity. Efforts were made to ensure that all AI-derived findings were transparently reported and critically assessed against established research standards to prevent the uncritical acceptance of AI-generated conclusions.

Conclusion: The case study demonstrated that while GenAI tools such as ChatGPT can significantly enhance the efficiency of data analysis in qualitative research, they must be used judiciously and in conjunction with human oversight. The integration of AI provides a valuable tool for expanding the analytical capabilities of researchers but requires careful management to avoid compromising the research's quality and ethical standards. This case highlights the complementary relationship between GenAI and human researchers, advocating for a balanced approach that leverages the strengths of both to achieve richer and more reliable research outcomes.

This practical example emphasises the importance of using AI as a supportive tool in qualitative research, offering a model for other researchers interested in exploring the integration of AI technologies into their analytical frameworks.

- Critical Discussion: Complementing Human Insight with AI in Qualitative Research

Table 6.8 provides a clear and concise comparison between AI-driven and human-driven analyses, highlighting both the strengths and limitations of each approach in qualitative research. It emphasises the need for a balanced methodology that harnesses the efficiency and capabilities of AI while leveraging human expertise for depth, accuracy, and ethical consideration. In other words, while AI excels in rapidly processing large volumes of data, identifying initial patterns, and offering broad overviews, it lacks the depth and contextual understanding inherent in human analysis. Human researchers can interpret complex social and cultural nuances, thereby provide essential insights that AI alone cannot. The integration of AI in qualitative research, therefore, should not seek to replace human judgment but to complement it. This synergy allows for a more comprehensive exploration of data, where AI's efficiency enhances human interpretative depth, ensuring research remains robust, ethically sound, and deeply informed. This balanced approach maximises the strengths of both AI and human capabilities, reinforcing the integrity and richness of qualitative analyses.

4 Conclusion

In the exploration of AI's role within research and analysis, a central theme has emerged: AI serves as an assistant, not a replacement for human intellect and creativity, aligning well with existing studies (e.g., Alqahtani et al., 2023; Oh et al., 2018; Zawacki-Richter et al., 2019). While AI can generate ideas and interpretations, the ultimate responsibility for any decision-making must rest with human researchers. Our case studies reaffirm the principle that while AI is an invaluable tool in the research arsenal, it is the human element that ultimately guides its use

Table 6.8 A comparison between AI-driven and human-driven analyses

Aspect	AI-driven analysis	Human-driven analysis
Approach to data	Handles large volumes efficiently; good for initial coding and categorisation.	More personalised and context-aware; excels in deep interpretative work.
Speed	Rapid processing; can analyse extensive datasets quickly.	Slower, as it involves meticulous scrutiny and consideration of context.
Accuracy	May generate 'hallucinations' or errors not present in data.	Typically more accurate due to contextual understanding and critical thinking.
Depth of insight	Identifies broad patterns and trends efficiently.	Provides deep insights, understanding subtleties and complexities of data.
Bias	Potentially perpetuates existing biases in training data.	Subject to personal biases, but can consciously mitigate them through critical thinking.
Interpretation	Limited to patterns it has been programmed to recognise; lacks understanding of deeper meanings.	Interprets data within theoretical and cultural contexts; able to understand sarcasm, humour, etc.
Interaction with data	Generally static; follows predefined algorithms for analysis.	Dynamic; adapts methods based on ongoing data analysis and evolving research questions.
Transparency	'Black box' nature can obscure how conclusions are derived.	Process and decisions are more traceable and explainable.
Ethical considerations	Requires careful oversight to ensure ethical use of data and AI capabilities.	Inherently involves ethical deliberation and consideration throughout the research process.
Replicability	Can be difficult due to stochastic nature and constant updates to AI models.	Easier to replicate if detailed methodological steps are followed.

and shapes its impact. Moreover, ensuring that AI does not perpetuate existing biases or introduce new biases into the research process is also important.

The application of AI in data interpretation is particularly pronounced in *interdisciplinary research*, where the breadth of knowledge required often stretches beyond the expertise of any single researcher. For instance, AI's ability to rapidly process and interpret data from diverse fields can bridge the gap between different areas of expertise.

References

Anthony, C., Bechky, B. A., & Fayard, A. L. (2023). "Collaborating" with AI: Taking a system view to explore the future of work. *Organization Science, 34*(5), 1672–1694.

Alqahtani, T., Badreldin, H. A., Alrashed, M., Alshaya, A. I., Alghamdi, S. S., bin Saleh, K., Albekairy, A. M., et al. (2023). The emergent role of artificial intelligence, natural learning processing, and large language models in higher education and research. *Research in Social and Administrative Pharmacy, 19*(8), 1236–1242.

Bearman, M., Ryan, J., & Ajjawi, R. (2023). Discourses of artificial intelligence in higher education: A critical literature review. *Higher Education, 86*(2), 369–385.

Bloom, B. S. (1956). *Taxonomy of educational objectives*.

Brożek, B., Furman, M., Jakubiec, M., & Kucharzyk, B. (2024). The black box problem revisited. Real and imaginary challenges for automated legal decision making. *Artificial Intelligence and Law, 32*(2), 427–440.

Díaz-Rodríguez, N., Del Ser, J., Coeckelbergh, M., de Prado, M. L., Herrera-Viedma, E., & Herrera, F. (2023). Connecting the dots in trustworthy artificial intelligence: From AI principles, ethics, and key requirements to responsible AI systems and regulation. *Information Fusion, 99*, 101896.

Duan, Y., Edwards, J. S., & Dwivedi, Y. K. (2019). Artificial intelligence for decision making in the era of big data–evolution, challenges and research agenda. *International Journal of Information Management, 48*, 63–71.

Durán, J. M., & Jongsma, K. R. (2021). Who is afraid of black box algorithms? On the epistemological and ethical basis of trust in medical AI. *Journal of Medical Ethics, 47*(5), 329–335.

Dwivedi, Y. K., Sharma, A., Rana, N. P., Giannakis, M., Goel, P., & Dutot, V. (2023). Evolution of artificial intelligence research in technological forecasting and social change: Research topics, trends, and future directions. *Technological Forecasting and Social Change, 192*, 122579.

Ekundayo, T., Khan, Z., & Nuzhat, S. (2024). Evaluating the influence of artificial intelligence on scholarly research: A study focused on academics. *Human Behavior and Emerging Technologies, 2024*(1).

Essien, A., Bukoye, O. T., O'Dea, X., & Kremantzis, M. (2024). The influence of AI text generators on critical thinking skills in UK business schools. *Studies in Higher Education, 49*, 1–18.

Fabris, A., Baranowska, N., Dennis, M. J., Graus, D., Hacker, P., Saldivar, J., Zuiderveen Borgesius, F., & Biega, A. J. (2023). *Fairness and bias in algorithmic hiring: A multidisciplinary survey*. ACM Transactions on Intelligent Systems and Technology.

Forehand, M. (2010). Bloom's taxonomy. *Emerging Perspectives on Learning, Teaching, and Technology, 41*(4), 47–56.

Gerlich, M., Elsayed, W., & Sokolovskiy, K. (2023). Artificial intelligence as toolset for analysis of public opinion and social interaction in marketing: Identification of micro and nano influencers. *Frontiers in Communication, 8*, 1075654.

Guo, K., Wu, M., Soo, Z., Yang, Y., Zhang, Y., Zhang, Q., Lin, H., Grosser, M., Venter, D., Zhang, G., & Lu, J. (2023). Artificial intelligence-driven biomedical genomics. *Knowledge-Based Systems, 279*, 110937.

Igbokwe, I. C. (2023). Application of artificial intelligence (AI) in educational management. *International Journal of Scientific and Research Publications, 13*(3), 300–307.

Oh, C., Song, J., Choi, J., Kim, S., Lee, S., & Suh, B. (2018). I lead, you help but only with enough details: Understanding user experience of co-creation with artificial intelligence. In *Proceedings of the 2018 CHI Conference on Human Factors in Computing Systems* (pp. 1–13).

Perkins, M., & Roe, J. (2024). The use of generative AI in qualitative analysis: Inductive thematic analysis with ChatGPT. *Journal of Applied Learning and Teaching, 7*(1).

Rolnick, D., Donti, P. L., Kaack, L. H., Kochanski, K., Lacoste, A., Sankaran, K., Ross, A. S., Milojevic-Dupont, N., Jaques, N., Waldman-Brown, A., & Luccioni, A. S. (2022). Tackling climate change with machine learning. *ACM Computing Surveys (CSUR), 55*(2), 1–96.

Sison, A. J. G., Daza, M. T., Gozalo-Brizuela, R., & Garrido-Merchán, E. C. (2024). ChatGPT: More than a "weapon of mass deception" ethical challenges and responses from the human-centered artificial intelligence (HCAI) perspective. *International Journal of Human–Computer Interaction, 40*(17), 4853–4872.

Vilhekar, R. S., & Rawekar, A. (2024). Artificial intelligence in genetics. *Cureus, 16*(1).

Vössing, M., Kühl, N., Lind, M., & Satzger, G. (2022). Designing transparency for effective human-AI collaboration. *Information Systems Frontiers, 24*(3), 877–895.

Zawacki-Richter, O., Marín, V. I., Bond, M., & Gouverneur, F. (2019). Systematic review of research on artificial intelligence applications in higher education–where are the educators? *International Journal of Educational Technology in Higher Education, 16*(1), 1–27.

Chapter 7
AI-Assisted Writing and Editing

Lilian Schofield and Xue Zhou

Abstract This chapter examines the transformative role of Artificial Intelligence (AI) in academic writing and editing, spotlighting Generative AI (GenAI) as a pivotal tool in modern academic practices. This chapter provides a comprehensive overview of how these AI tools are employed to streamline the writing process, from the initial stages of idea generation to the final stages of proofreading and citation management. In addition, we discuss the practical applications of AI in structuring coherent and logical academic narratives and its effectiveness in improving language clarity and coherence for both native and non-native English speakers. The chapter also critically addresses the ethical considerations and potential pitfalls of relying heavily on AI for academic writing. It highlights the importance of maintaining a balance between AI assistance and the writer's original insight to preserve academic integrity and intellectual authenticity. By integrating GenAI tools thoughtfully, academic writers can enhance their productivity while upholding the standards of scholarly writing. The discussion aims to equip readers with knowledge of how to effectively and ethically integrate AI tools into their writing practices, ensuring that these technologies serve as aids rather than substitutes for human intellect and creativity.

L. Schofield (✉)
Queen Mary University of London, London, UK
e-mail: l.schofield@qmul.ac.uk

X. Zhou
University of Leicester, Leicester, UK
e-mail: xue.zhou@le.ac.uk

X. Zhou, H. Al-Samarraie (eds.), *Institutional guide to using AI for research*,
Advances in Artificial Intelligence in Education 2,
https://doi.org/10.1007/978-3-031-94809-1_7

1 Introduction: The Growing Role of AI in Academic Writing

Academic writing is a process that closely intertwines with research and plays an active role in knowledge creation and the outputs include research papers, journal articles, dissertations and theses, literature reviews, case studies, monographs, conference papers, textbooks, and scholarly books (Badley, 2009). It enables writers to communicate their ideas effectively, articulate their thoughts clearly, and reach their various audiences through a structured approach that emphasises clarity, coherence, critical analysis, precise language, evidence-based arguments, proper citation of sources, structured organisation, audience awareness, objective tone, and logical reasoning. (Khalifa & Albadawy, 2024; Kurniati & Fithriani, 2022; Song & Song, 2023). The purpose of academic writing varies and includes disseminating new knowledge and research findings, critically examining existing theories and practices, developing and refining arguments within a field, demonstrating understanding and expertise, and contributing to ongoing scholarly conversations.

There are various key aspects of writing that writers struggle with, such as having ideas about how to start writing, as writers struggle to come up with ideas or expand on existing ideas. This process can be particularly challenging when faced with new areas for exploration, complex topics or fields where innovation is crucial. Structuring a paper can present another challenge for writers, as organising thoughts into a coherent, logical flow that effectively communicates the intended message can be a complex task. Writers can struggle with deciding how to organise their arguments. In addition, exploring literature can be daunting due to the sheer volume of existing research which can be overwhelming. Editing and proofreading is a process of refining and polishing a piece of writing and this can be time-consuming, and for non-native English language speakers, this challenge can be further complicated. Citing properly is yet another important aspect of academic writing that includes various citation styles, such as APA, MLA, or Chicago. Part of academic writing is that sources must be correctly attributed, and the chosen citation style consistently applied throughout the writing. However, with the introduction of AI, these challenges can now be automated by the different AI-tools.

In academic writing, AI tools are proving to be game-changers with Generative AI (GenAI) tools, such as ChatGPT used for educational learning and academic writing (Cotton et al., 2023; Tang et al., 2024). This has made generative AI tools useful for various applications in education, teaching, and writing (Cotton et al., 2023; Nguyen et al., 2024). Large Language Model (LLM), which includes ChatGPT (OpenAI), Bard (Google), and Claude (Anthropic), are useful AI tools that aid not only in exploring and structuring literature reviews but also in writing (Tang et al., 2024). Tools such as Grammarly, which employs advanced natural language processing algorithms, can be used whilst writing the literature review, check grammar, style, and tone and even offer suggestions to improve clarity and coherence (Ooi et al., 2023). Writers can use these tools to generate drafts, refine writing styles, and check for grammatical errors. However, it is important to note that the output is also based on the prompts and users' prior knowledge. As highlighted by Essien et al. (2024), coming up with relevant prompts that effectively align with questions the writers want an answer to can be challenging. With relevant prompting, these tools can help generate outlines, suggest phrasings, or even draft sections of papers (Malinka et al., 2023; Ooi et al., 2023).

Although technology-assisted writing is not a new concept, the recent introduction of ChatGPT, developed by OpenAI in 2021 signify a significant evolution in this area (Cotton et al., 2023; Nguyen et al., 2024). With recent advancements, these AI technologies can generate content, produce text closely resembling human writing, support writing based on prompts, summarise articles, and make the writing process efficient (Huang & Tan, 2023; Jarrah et al., 2023; Kasneci et al., 2023; Nguyen et al., 2024). For example, ChatGPT can help writers brainstorm ideas, draft introductions, or summarise long texts, even drafting entire sections of papers and making the writing process more efficient (Wang, 2024). AI tools like QuillBot and Grammarly can paraphrase sentences, helping writers rephrase and refine their academic content when properly prompted. They can also include sentiment analysis, which understands the tone and sentiment of a piece of writing. Users can prompt the tools to make the sentence more professional or enthusiastic. AI tools like Zotero and Mendeley help manage citations and references, automatically formatting them according to various academic styles. They can import references from databases, generate bibliographies, and integrate with word processors. For example, Mendeley can automatically extract metadata from PDFs, making it easier to organise and cite sources accurately. Another aspect of academic writing that can benefit from AI tools is the generation of ideas and pre-reading. ChatGPT, Elicit, and other AI tools can help writers quickly grasp a text's main ideas and arguments without reading the whole document. These tools use natural language processing and machine learning as noted above to summarise, extract, and highlight key information from a text, such as a journal paper, abstract, methodology and conclusions. They can also provide additional features, such as citations, keywords, links, and questions, to help writers further explore the topic and find relevant sources. These tools can save time and energy in the pre-writing stage, as writers can easily scan through multiple texts and identify the ones that are most relevant and useful for their own writing (Khalifa & Albadawy, 2024). AI tools are helpful in developing inspiration and guidance for developing one's own ideas and arguments while avoiding plagiarism and ensuring originality. However, these AI tools are not substitutes for reading the original texts, as they may omit or misinterpret some important details or nuances. Writers should always verify the accuracy and validity of the information provided by AI tools and read the original texts carefully and critically before incorporating them into their own writing. This way, they can ensure that their writing is well-informed, well-supported, and well-written.

This chapter provides some practical suggestions and examples of ways to use generative AI tools to make writing more streamlined and efficient, especially in overcoming some of the challenges mentioned above. However, it is important to note that these tools should not write or replace users writing but support them in the writing process. Generative AI tools discussed in this chapter provide examples of how the tools can support writing, especially in the areas of editing, structure, literature search and categorisation and writing ideas. The chapter does not focus on promoting generative AI tools nor is it about the operational use of GenAI, as these tools evolve with more tools introduced or existing features advanced with innovations and updated. This chapter is about developing the thinking about the efficient use of GenAI for academic writing and how each tool supports aspects of academic writing. The chapter also seeks to show what writers should think about as they utilise these tools.

2 Considerations when Using AI in Writing

When using GenAI, there are several considerations to be aware of because without considering how to utilise these tools and tailor them to one's specific needs, writing may become more frustrating than helpful or efficient. Having a good understanding of one's writing profile is important in selecting the right tools to complement and enhance the writing process. Therefore, it is essential to recognise that not every GenAI writing tool available will be suitable for every type of writing; the key is to identify those that align with one's own specific style of writing, the requirements of the academic paper and the needs, and objectives of the paper. Further, not every tool will suit every writer. For instance, if a user's approach to writing is highly structured, they may benefit from tools that assist with outlining and mapping ideas in a clear and structured manner rather than complicated to use tools or using multiple tools at the same time. Though several GenAI tools can perform similar functions, such as Jenni AI, Perplexity, ChatGPT and Claude AI, writers do not have to use all in writing one piece of writing as using several simultaneously may not be more efficient.

It is important to be clear about how these tools can be integrated into one's writing style and approach, knowing that the effectiveness of a writing tool also depends on how well it integrates into an existing way of writing. Tools that are complex to use and require further learning or significant changes to a user's routine might disrupt their creative process rather than facilitate it. Therefore, it is important to trial tools and evaluate their impact on productivity and writing quality before fully integrating them into practice. Further, consider the efficiency of using only relevant generative AI tools for writing rather than several. The aim is not to use as many AI tools as possible but to use them effectively and efficiently to support writing and approach to writing. The AI tools should not write for users but should help streamline the writing process.

3 GenAI in Academic Writing

3.1 GenAI Tools that Support Prereading

Writing requires knowledge of relevant literature to determine what has already been written in the field and where gaps exist. Reading enables writers to develop their thoughts and support their arguments with literature. In the past, a writer would search through vast amounts of databases and journals to find relevant literature; however, with AI tools now introduced to streamline the writing process AI tools like Elicit, Consensus, Research Rabbit, LitMaps, etc., can make the process more streamlined with writers able to use the tools for literature searches, systematically organising the research papers and extracting critical information. An AI tool that can aid in categorising and organising literature is Elicit. The AI tool can also be used for literature searching and prereading to have an overview of the literature and go through relevant sections. It can summarise and give concise overviews based on the prompt given. Users can specify the information they need, and information will be neatly presented and organised into tables, aiding the comparison and analysis of data across various papers.

Consensus AI is another academic search engine that provides a quick insight into the main findings of research papers. Its database covers various disciplines and provides access to various academic papers. It can also provide personalised recommendations based on prior searches. Writers can search for literature and be presented with various literature and a link to read the papers. It also makes it easy to cite the sources. Connected Papers is another academic AI tool designed to help identify and explore academic papers. Writers can input the search topics they would like to explore, and it responds to the search by visually providing connected related papers, where each node corresponds to a paper. Edges denote connections based on citations and thematic similarities (Fig. 7.1).

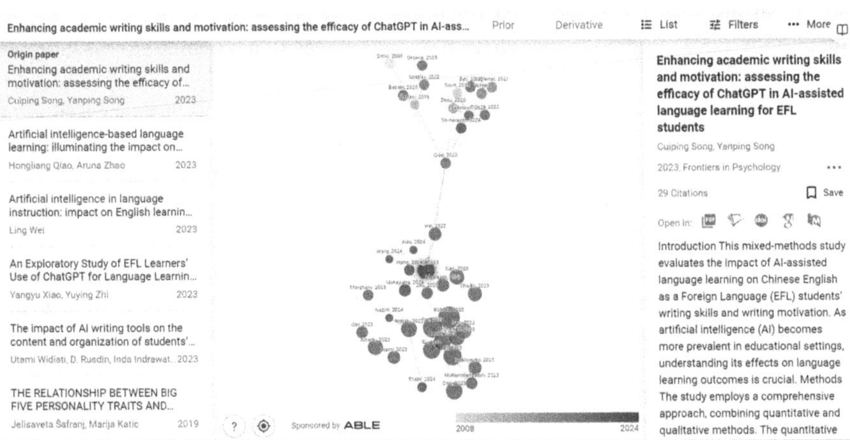

Fig. 7.1 Connected papers

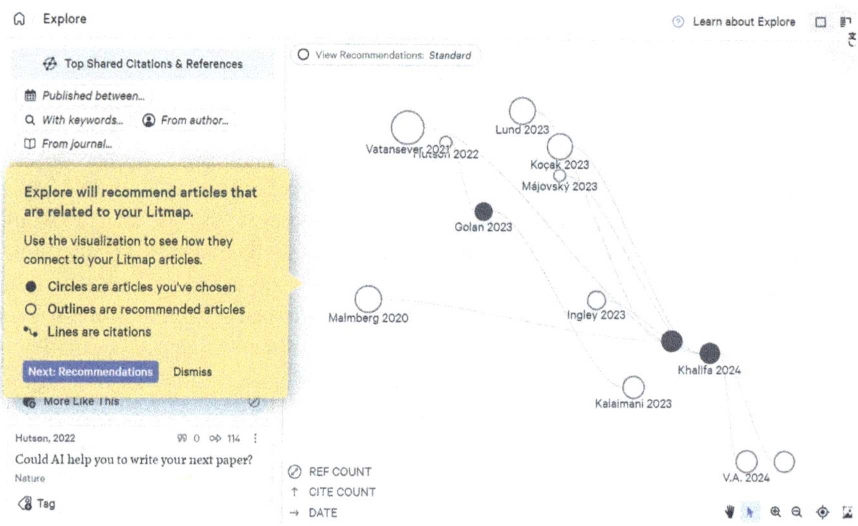

Fig. 7.2 Litmaps

Writers can click on nodes to find more information about each paper, including titles, authors, abstracts, and publication years. Additionally, writers can read summaries of each paper to get an overview. This visual representation enables writers to easily identify relevant papers and understand their interconnections. This tool can be instrumental in streamlining the literature search process.

ResearchRabbit is another AI tool that streamlines the literature search process. The AI tool can map papers on a given topic visually. Search can be done by inserting a topic, author, abstract, and citations. Litmaps is a useful AI tool for writing literature reviews and can help streamline searching. Identifying relevant papers for a literature review or academic writing can be time-consuming for writers. Litmaps assists in discovering papers, visualising relationships between them, and sharing them with users' reference manager, making it easier to navigate the vast literature (Fig. 7.2).

The litmaps visual shows the topic searched for. The search process is quite straightforward. Click on "Create your first Litmap," and a search bar will appear, allowing users to search for papers using keywords, titles, author names, and more. For the above example, "AI for academic writing" was inserted in the search. Litmaps presented a list of related papers and their abstracts, and three relevant papers were selected before clicking "Create Litmap,". Litmaps then generated a visual map displaying my chosen papers. The dark circles are the articles selected. The outlines are recommended articles that are relevant when searching for relevant papers. Users can click on the 'more like this paper' to explore relevant literature. They can also click on the outlines in the visual to explore relevant literature. Finally, users can export the selected papers to their reference manager.

3.2 *Using AI for Structure of the Writing*

The structure of a paper is important in academic writing, especially for coherency and flow of argument. Generative AI tools like ChatGPT, Monica, Jenni AI, and Claude can help provide the initial structure or ideas for structure. By inputting a simple prompt, these AI tools can quickly generate a comprehensive structure that outlines key points and suggests a logical flow. This capability helps with efficiency, saving time and providing areas for refining arguments and enhancing content quality rather than getting stuck on the organisation. However, it is important to note that the structures provided by AI are starting points for idea generation and should be adapted to meet specific writing needs and personal or academic standards. Writers can modify these automated outlines to reflect better their unique voice, depth of knowledge, and the specific requirements of their audience or publication. This adaptive use of AI in writing ensures that the technology is a support tool, enhancing creativity and effectiveness without diminishing the writer's ownership and individuality. This approach not only streamlines the writing process but also enriches the content, making it more tailored and impactful.

Example 1: using Claude AI for structure

The prompt I instructed Claude AI

> L I am writing an essay on academic writing, please draft an outline for a 500-word essay on academic writing.

The result from Claude AI (Fig. 7.3).

The structure may differ from initial ideas; however, users can amend and adapt it to their needs. This step is only necessary if users want to check if the AI tool brings new ideas for structure. Users should have an idea of what their structure should be or follow the structure provided by tutors or journal submissions. It is not advisable to rely on AI tools without own thinking.

3.3 *Editing, Review, and Publishing Support*

After reading and gathering vast amounts of information, it is now time to put words into writing. Sometimes, this can be a challenge for writers. Khalifa and Albadawy's (2024) work highlights how generative AI can support editing, review, and publishing. Therefore, these tools can be used to draft initial ideas, providing a foundation that writers can build upon, particularly in the initial stages of idea generation and development. For instance, writers might utilise generative AI tools to capture their thoughts without the immediate pressure of structuring or refining them. Once these initial thoughts are recorded, they can be transferred to generative AI tools such as ChatGPT or Claude for further refinement. Writers can paste their raw thoughts into

Certainly. Here's an outline for a 500-word essay on academic writing:

I. Introduction (50 words)
A. Definition of academic writing
B. Importance in higher education and research
C. Thesis: Academic writing is a vital skill that requires mastery of specific conventions and techniques to effectively communicate scholarly ideas.

II. Key Characteristics of Academic Writing (150 words)
A. Formal tone and style
1. Third-person perspective
2. Precise language
B. Objectivity
1. Emphasis on evidence
2. Minimizing personal bias
C. Clarity and conciseness
D. Use of discipline-specific terminology
E. Logical organization and structure

III. The Academic Writing Process (150 words)
A. Pre-writing
1. Understanding the assignment
2. Research and outlining
B. Drafting
C. Revising and editing
D. Proofreading
E. Importance of citation and avoiding plagiarism

IV. Types of Academic Writing (100 words)
A. Essays (argumentative, expository, analytical)
B. Research papers
C. Dissertations and theses
D. Journal articles
E. Other types (e.g., lab reports, case studies)

V. Conclusion (50 words)
A. Recap of key points
B. Importance of developing academic writing skills
C. Encouragement for continuous improvement in academic writing

This outline provides a structured approach to discussing academic writing within the 500-word limit. Would you like me to explain any part of this outline further?

Fig. 7.3 Results from Claude AI

the generative AI tools and prompt them to 'clean up' the text, organise ideas, and enhance clarity without altering the original intent or content. As a preliminary editing tool, AI allows writers to view their ideas in a more polished form, providing a clearer direction for subsequent revisions and expansions. However, the key is that the AI tool remains an assistant, not the writer, and the work remains the author's own, preserving the originality and authenticity of their ideas while AI tools refine the raw ideas. This emphasises the distinction between AI as a facilitator of the writing process and the human author as the true originator of thought. It is also important to be aware that these tools can alter the original meaning or intended meaning of the author's words and ideas and it is always important to check that the text generated by AI is as intended and the original meaning not lost.

In addition to idea generation capture and development, generative AI tools can assist in drafting content. For instance, writers can input a rough outline or a series of bullet points into a generative AI tool, which can then expand on these points, providing more ideas that the writer can later refine. The AI-generated drafts serve as a starting point, enabling writers to focus on adding depth, nuance, and personal insights, thus enhancing the overall quality of the articles while reducing the time spent on initial drafting.

Example 2 Using Generative AI to Clean up Interview Transcripts or Recorded Ideas

Generative AI can clean up transcripts, including users own recorded thoughts. As is often the case with spoken language, the raw interview transcripts can be rough, capturing many filler words, repetitions, and instances where the speakers' words are muddled or unclear. Furthermore, speaking in a natural, conversational style often leads to fragmented sentences and overlapping dialogue, which need to be refined for clarity and coherence. Before pasting transcripts to ChatGPT to help refine the content, remove all personal and sensitive information from the transcripts to ensure privacy and confidentiality. After pasting the raw transcript into ChatGPT, prompt it to remove duplications and filler words such as "um," "uh," "like," and "you know," as well as unnecessary repetitions that are typical in spontaneous speech without changing the interview or introducing new vocabulary. Once this preliminary cleanup is done, prompt ChatGPT to polish the transcript, transforming it into a more coherent and readable format without changing the words. Using ChatGPT or even Claude in this way, writers can streamline the process of transcribing and editing the interviews.

3.4 Using AI for Citation of Sources

The citation of sources is fundamental to maintaining academic integrity and is an essential aspect of academic writing. Proper citation not only gives credit to the original authors but also allows readers to trace the development of ideas and research findings. However, managing citations can be complex and

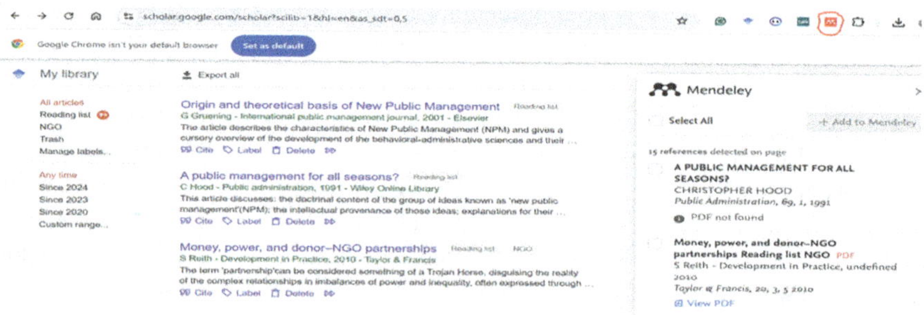

Fig. 7.4 Exporting from Google Scholar to Mendeley

time-consuming, particularly when dealing with multiple source types and citation styles. This is where AI tools like Mendeley and Zotero come into play in facilitating reference management. Mendeley and Zotero significantly simplify the citation process by automating the creation of in-text citations and reference lists according to academic standards such as APA, MLA, or Chicago styles. These tools not only save time but also minimise the risk of errors, which can be crucial for academic credibility. Automatically formatting citations and keeping track of all sources help writers focus more on the content of their work rather than the intricacies of citation rules. Additionally, their ability to sync across different AI tools enhances efficiency, making integrating into other AI tools and managing references easier. Some of the tools highlighted above demonstrate how tools like Zotero and Mendeley are integrated into them for efficient citation.

Exporting from Google Scholar to Mendeley.

As shown in Fig. 7.4, If a library is already available, the Mendeley Web Importer extension can be imported. To do this, click on the Mendeley extension as highlighted above to view the readings in the Google Scholar Library and select those to be added.

3.5 *AI-Assisted Proofreading*

AI-assisted proofreading can be a useful tool to enhance the quality and readability of academic writing. However, it is important to be aware of the ethical implications and limitations of using such tools, especially concerning academic integrity and originality. There are certain areas where the usage of generative AI is still contested, and many academics argue that it is unacceptable to use a tool to rewrite the original text or generate new text or material. Even journal reviewing is being impacted by AI with some journal publishers having banned the use of AI, and some, like Springer-Nature, have a guideline that researchers report their use of LLM tools in the methods or acknowledgements section (Eggmann et al., 2023). Some publishers and institutions allow students to use AI for proofreading, but it is essential to check institutional and publisher guidelines. For example, AI tools such as ChatGPT or Grammarly can be used to identify and correct spelling, grammar, punctuation, formatting, and presentation issues. However, it is also important to remember that there are issues with intellectual property. Therefore, AI-assisted proofreading requires careful judgment and critical evaluation of the suggested changes and additions. Users should always maintain their own voice and style and avoid plagiarism and fabrication of information, as is known with some AI-generated information.

Example 3 Using AI for Proofreading
AI tools like Grammarly, ChatGPT, Quilbot and many others can be used to proofread and polish writing. However, before using the tools for proofreading or editing, users should consider starting by thoroughly reviewing and editing their work independently. This ensures that their core ideas and arguments remain intact and that they have addressed the main structural and content issues. To check for grammar issues, users should run their document through an AI-powered grammar and spelling checker, such as Grammarly, Quilbot or ProWritingAid. These tools can identify basic errors that might have been missed, including punctuation and typographical errors. Some may have these tools downloaded and embedded into their writing tools like MS Word, and these AI tools automatically highlight grammar issues. AI tools like ChatGPT can be used to check specific problematic paragraphs or sentences. For example, users could input: ***"Please suggest ways to improve the clarity and concision of the following paragraph without changing its argument or sentence structure."*** Some writing requires consistency in the use of certain terminologies. AI tools such as ChatGPT can help with it. To do this, prompt ChatGPT with the following prompt after giving it the text to check: "Please review this text for consistent use of technical terms and highlight any inconsistencies." In editing, users may also want to check for consistency with citations and references. To check for this, use any of the AI tools mentioned to double-check citations and references. Users may use this prompt on ChatGPT: ***"Please verify if the following citations are in the correct APA (or other relevant style) format.".*** The paid version of Grammarly also highlights the consistency in citations.

After using generative AI tools for editing and grammar checks, users should thoroughly review the document and ensure that the AI-suggested changes are consistent with their intended meaning and academic tone. If possible, users should ask a colleague or peer to review their work as well. This human perspective can identify nuances that AI might overlook.

Remember, while using AI for editing and proofreading:

- Always check the guidelines. If in doubt, check with module leaders, publishers or anyone in charge of the guidelines.
- Do not just accept changes without critically evaluating the generated suggestions.
- AI should not write for the user. The tools should only be used for assistance.
- Users should check the guidelines and be transparent about using AI tools as required by their institution or publishers.

4 Ethical Considerations and Issues Related to the Use of AI in Academic Writing (e.g., Authorship, Intellectual Property, Bias, and Transparency)

Though other chapters in this book touch on ethical implications, it is important to highlight them regarding academic writing using AI. Integrating AI tools like ChatGPT in writing, particularly in academic contexts, raises important concerns regarding academic integrity and the authenticity of AI-generated content (Chan, 2023). One primary concern is associated with academic integrity (O'Dea, 2024). AI tools can be used in various aspects of writing, including structuring, coherence, grammar, and restructuring of sentences (Song & Song, 2023). There are still debates about what constitutes academic misconduct and whether using AI for grammar checks falls under academic misconduct. It is critical to follow the guidelines of the journal or assignments on whether the use of AI is permitted and, if so, to what extent. This goes back to the argument that writing should be the original thought of the writer and attribution (Tang et al., 2024).

The growing reliance on AI in academic writing raises critical concerns about the authenticity of scholarly work and the potential erosion of essential skills. There are arguments that there is a risk that over-dependence on AI tools could lead to a decline in critical thinking abilities, originality, and the nuanced understanding that comes from deep engagement with source material. Further, the use of AI in academic writing raises ethical concerns regarding authorship, intellectual property, and academic integrity (Lodge et al., 2023). The Hallucination effect (Essien et al., 2024), bias, fairness, accountability, and transparency (Ooi et al., 2023) are also a concern. Additionally, the accessibility of AI writing tools could exacerbate existing inequalities in academia, disadvantaging those without access to advanced versions and potentially widening the gap between those who can afford the subscription-based version and those who cannot (Zhou and Schofield). However, educators and institutions are developing guidelines to balance the use of AI in teaching, research and learning. For example, some universities are already implementing policies to guide the ethical use of AI in academic writing, ensuring that students use these tools responsibly (Schofield & Zhang, 2024).

Another issue is the over-reliance on AI for tasks traditionally reserved for human cognitive processes, such as critical thinking and analysis, could be problematic. AI tools can proficiently produce content that superficially meets assignment criteria but may lack the depth and originality that come from human critical thinking. Academic work thrives on unique insights and personal engagement with material, aspects that AI currently cannot replicate. Educators and students must be cautious not to substitute AI for human judgment and intellectual effort. Furthermore, heavy reliance on AI tools could undermine the development of writing skills. Regular use of AI for drafting, revising, and editing can potentially deter students from practising these essential skills.

5 Best Practices for Responsible and Ethical Use of AI in Academic Writing

To ensure the responsible and ethical use of AI tools in academic writing, it is crucial to adhere to a set of best practices that uphold academic integrity. The following guidelines are essential considerations:

- Transparency is also crucial when using AI-generated content in academic settings and writers must explicitly disclose the use of AI, whether for idea generation, text drafting, or editing purposes.
- It is crucial to use AI as a support tool rather than a substitute for original thought and research. While AI can assist in generating initial drafts and providing suggestions, the core ideas, arguments, and critical analysis should come from the writer's own understanding and thinking.
- Critical engagement with AI-generated content is essential, and generated content must be rigorously evaluated for relevance and accuracy of information generated by AI tools. This process involves carefully reviewing AI-generated outputs, cross-referencing with academic sources, and making necessary adjustments to align the content with academic standards and the writer's unique voice.
- Adherence to institutional and publication-specific ethical guidelines regarding AI use in academic writing is crucial. These guidelines, established by educational institutions, journal publishers, or other academic outlets, serve to maintain consistent academic standards and ensure fairness in evaluation processes.
- Proper citation practices must extend to include AI tools that contributed to the work. This approach respects intellectual property rights and maintains the integrity of academic contributions. By acknowledging AI assistance, authors provide a complete picture of their research methodology and writing process, allowing readers and evaluators to fully understand the work's context and development.

6 Conclusion

As AI continues to evolve, its role in academic writing will likely expand with guidelines for using AI in writing that are more universal and clearer on how to use AI tools for writing support. The key to using these tools ethically and responsibly in writing will be striking a balance between leveraging the efficiency and capabilities of AI tools while maintaining human insight, creativity, and critical thinking that are at the heart of academic inquiry. While AI tools like ChatGPT offer transformative capabilities in streamlining and enhancing the academic writing process, their use in academic writing must be carefully considered to maintain academic integrity and ethical standards. Transparency, originality checks, and critical engagement are paramount in ensuring these tools serve as aids rather than substitutes for human intellect and effort. It is also important to note that with the advancement of technology, some of the tools available now may be redundant later or have some of their features changed. Also, some of the AI tool's most advanced features can be accessed on a subscription basis. Further, different tools have different functions, and users would have to be clear about their goals in writing and how the tools can be leveraged to support their writing. This is to prevent using various AI tools that perform similar functions at the same time that they end up being confused about how to use them and end up frustrated, not being able to achieve their writing goals and not saving time. The key for users is to know who they are as a researcher and writers and what tools suit their writing style and research. Further, think of efficiency and knowledge of usage when thinking of which tools fit their profile.

References

Badley, G. (2009). Academic writing as shaping and re-shaping. *Teaching in Higher Education, 14*(2), 209–219. https://doi.org/10.1080/13562510902757294

Chan, C. K. Y. A. (2023). Comprehensive AI policy education framework for university teaching and learning. *International Journal of Educational Technology in Higher Education, 20*, 38. https://doi.org/10.1186/s41239-023-00408-3

Cotton, D. R. E., Cotton, P. A., & Shipway, J. R. (2023). Chatting and cheating: Ensuring academic integrity in the era of ChatGPT. *Innovations in Education and Teaching International, 61*(2), 228–239. https://doi.org/10.1080/14703297.2023.2190148

Eggmann, F., Weiger, R., Zitzmann, N. U., & Blatz, M. B. (2023). Implications of large language models such as ChatGPT for dental medicine. *Journal of Esthetic and Restorative Dentistry, 35*(7), 1098–1102. https://doi.org/10.1111/jerd.13046

Essien, A., Bukoye, O. T., O'Dea, X., & Kremantzis, M. (2024). The influence of AI text generators on critical thinking skills in UK business schools. *Studies in Higher Education, 1–18*, 865–882. https://doi.org/10.1080/03075079.2024.2316881

Huang, J., & Tan, M. (2023). The role of ChatGPT in scientific communication: Writing better scientific review articles. *American Journal of Cancer Research, 13*(4), 1148. www.ajcr.us/ ISSN:2156-6976/ajcr0150104.

Jarrah, A. M., Wardat, Y., & Fidalgo, P. (2023). Using ChatGPT in academic writing is (not) a form of plagiarism: What does the literature say? *Online Journal of Communication and Media Technologies, 13*(4), e202346. https://doi.org/10.30935/ojcmt/13572

Kasneci, E., Seßler, K., Küchemann, S., Bannert, M., Dementieva, D., Fischer, F., Gasser, U., Groh, G., Günnemann, S., Hüllermeier, E., & Krusche, S. (2023). ChatGPT for good? On opportunities and challenges of large language models for education. *Learning and Individual Differences, 103*, 102274. https://doi.org/10.1016/j.lindif.2023.102274

Khalifa, M., & Albadawy, M. (2024). Using artificial intelligence in academic writing and research: An essential productivity tool. *Computer Methods and Programs in Biomedicine Update, 100145*. https://doi.org/10.1016/j.cmpbup.2024.100145

Kurniati, E. Y., & Fithriani, R. (2022). Post-graduate students' perceptions of Quillbot utilization in English academic writing class. *Journal of English Language Teaching and Linguistics, 7*(3), 437–451. https://doi.org/10.21462/jeltl.v7i3.852

Lodge, J. M., de Barba, P., & Broadbent, J. (2023). Learning with generative artificial intelligence within a network of co-regulation. *Journal of University Teaching & Learning Practice, 20*(7). https://doi.org/10.53761/1.20.7.02

Malinka, K., Peresíni, M., Firc, A., Hujnák, O., & Janus, F. (2023, June). On the educational impact of chatgpt: Is artificial intelligence ready to obtain a university degree?. In *Proceedings of the 2023 Conference on Innovation and Technology in Computer Science Education* V. 1 (pp. 47–53). https://doi.org/10.1145/3587102.3588827

Nguyen, A., Hong, Y., Dang, B., & Huang, X. (2024). Human-AI collaboration patterns in AI-assisted academic writing. *Studies in Higher Education, 49*(5), 847–864. https://doi.org/10.1080/03075079.2024.2323593

O'Dea, X. (2024). Generative AI: Is it a paradigm shift for higher education? *Studies in Higher Education, 49*(5), 811–816. https://doi.org/10.1080/03075079.2024.2332944

Ooi, K. B., Tan, G. W. H., Al-Emran, M., Al-Sharafi, M. A., Capatina, A., Chakraborty, A., Dwivedi, Y. K., Huang, T. L., Kar, A. K., Lee, V. H., & Loh, X. M. (2023). The potential of generative artificial intelligence across disciplines: Perspectives and future directions. *Journal of Computer Information Systems, 65*, 1–32. https://doi.org/10.1080/08874417.2023.2261010

Schofield, L. & Zhang, J. J. (2024). Spotlight on the inclusion process in developing AI guidance and policy. Available: https://srheblog.com/?p=4921&preview=1&_ppp=8db0186ba3

Song, C., & Song, Y. (2023). Enhancing academic writing skills and motivation: Assessing the efficacy of ChatGPT in AI-assisted language learning for EFL students. *Frontiers in Psychology, 14*, 1260843. https://www.frontiersin.org/journals/psychology/articles/10.3389/fpsyg.2023.1260843/full Accessed 01/08/2024

Tang, A., Li, K. K., Kwok, K. O., Cao, L., Luong, S., & Tam, W. (2024). The importance of transparency: Declaring the use of generative artificial intelligence (AI) in academic writing. *Journal of Nursing Scholarship, 56*(2), 314–318. https://doi.org/10.1111/jnu.12938

Wang, C. (2024). Exploring students' generative AI-assisted writing processes: Perceptions and experiences from native and nonnative English speakers. *Tech Know Learn.* https://doi.org/10.1007/s10758-024-09744-3

Part III
Ethical Considerations in AI-Enabled Research

Chapter 8
Ethical Implications and Governance of AI in Research

Lucy Caton and Yujeong Hwang

Abstract This chapter examines the ethical implications of using Generative AI in postgraduate research, with a focus on accountability, transparency, and generative AI governance. It reviews the current AI policy landscape in the UK and beyond, contextualising accountability and transparency within higher education, and evaluates existing frameworks, proposing additional measures to enhance these aspects. This includes ensuring that generative AI-driven research upholds ethical standards and continues to make both impactful and authentic contributions to academia. Furthermore, the chapter will explore the role of higher education institutional policies in guiding postgraduate students' responsible use of generative AI tools for research, highlighting the importance of clear, accessible guidelines and support systems. The aim is to raise awareness among postgraduate students about current generative AI policy discourse and developments for academic pursuits. Additionally, the empirical case study, offers a way to empower postgraduate students to address biases in their work whilst showcasing an example of ethical and transparent generative AI institutional governance. By analysing existing policy frameworks within the wider sector, we aim to critique yet further cultivate responsible and innovative research environments.

Keywords Postgraduate research · Policy · Generative AI · Transparency · Governance · Accountability

L. Caton (✉) · Y. Hwang
The University of Greater Manchester, Bolton, UK
e-mail: L.Caton@bolton.ac.uk

X. Zhou, H. Al-Samarraie (eds.), *Institutional guide to using AI for research*, Advances in Artificial Intelligence in Education 2,
https://doi.org/10.1007/978-3-031-94809-1_8

1 Introduction

Artificial Intelligence (AI) is slowly becoming an integral part of academic research, the introduction of AI tools, especially generative AI, has brought about significant changes in how research is conducted, analysed, and disseminated. For clarity this chapter adopts the term "generative AI" to refer specifically to a subset of artificial intelligence that generates new content, such as text, images, music, or full conversations, based on its learned data. We delve into the ethical dimensions of accountability and transparency frameworks within postgraduate academic study, while also examining broader policy and governance dynamics that are shaping the way forward for higher education institutions. While generative AI offers considerable benefits to postgraduate students, enhancing productivity, creativity, and work quality through multidisciplinary perspectives (Dwivedi et al., 2021), it also presents challenges and constraints, including ethical, social, and epistemological concerns (Capraro et al., 2024; Nah et al., 2023; Selwyn, 2022;), as well as the necessity for human oversight, validation, and interpretation of AI-generated outputs. Notably, the ethical implications of authenticity and potential misuse of generated content raise pertinent questions about intellectual property rights, authorship attribution, and the risk of misinformation. Moreover, the opacity of many generative AI models complicates result validation, eroding trust in research findings (Liu & Jagadish, 2024), and the presence of biases in training data can perpetuate societal inequalities in generated outputs, necessitating vigilant consideration and mitigation strategies (Özer et al., 2024).

The growing significance of generative AI in academic research cannot be overstated. AI technologies, such as machine learning and natural language processing, have enabled researchers to handle complex datasets and perform sophisticated analyses that were previously unimaginable. For instance, generative AI can automate labour-intensive tasks like data collection and preliminary analysis, thereby enhancing the efficiency and accuracy of research processes (Pal, 2023; Juarez-Orozco et al., 2019). This automation not only saves time but also reduces human error, leading to more reliable and reproducible results (Burger et al., 2023; Neyedli et al., 2011).

Generative AI's capabilities extend beyond mere automation, for example, advanced machine learning algorithms can conduct intricate content analyses, identify emerging trends, and highlight gaps in existing research, thus facilitating deeper engagement with research questions (Nguyen-Trung et al., 2023). Generative AI's prowess in natural language processing has also revolutionised literature reviews, enabling the scanning and summarisation of vast amounts of literature, which helps researchers stay abreast of the latest developments in their fields (Müller et al., 2022).

The interdisciplinary potential of generative AI is another noteworthy aspect for postgraduate research. By analysing datasets from diverse academic fields, generative AI fosters novel interdisciplinary collaborations and projects, thereby enriching academic discourse and expanding the scope of research (Kox & Beretta, 2024). However, the integration of generative AI in research is not without its challenges.

Ethical concerns, particularly regarding accountability and transparency, have come to the forefront as generative AI systems become more prevalent in academic settings. Accountability in AI-driven research involves ensuring that generative AI systems are used responsibly and that their outputs can be traced back to their sources. This is crucial for maintaining the integrity of research and for addressing any potential biases or errors in AI-generated results (Chen et al., 2024; Memarian & Doleck, 2023). Transparency, on the other hand, requires that the processes and decisions made by generative AI systems are understandable and accessible to all stakeholders, including non-experts (Spiegelhalter, 2020). This is essential for building trust in generative AI systems and for ensuring that their use in research is ethical and justifiable.

Additionally, governments worldwide are becoming more involved in regulating the use of generative AI across the sectors, including academia. The European Union, the United States, United Kingdom and China have drafted national policies, as out lined in the coming section, aimed at governing the wider ethical use of AI technologies. These policy frameworks aim to balance innovation in AI technologies whilst safeguarding societal values and mitigating potential harms. The ongoing development of such regulatory measures will be critical as generative AI continues to play a transformative role in academia and beyond.

2 Navigating Bias and Equity in AI-Enhanced Postgraduate Research

Global policy development in higher education is influenced by both external and internal factors and with the rise of generative AI in research and development, these influences have become even more complex. As discussed above, generative AI has brought about innovative ways of conducting research that advances new techniques, methods and complex datasets (Smith et al., 2024), however these advancements also pose challenges to traditional research practices.

In response to these challenges, the higher education sector on an international level has been compelled to critically evaluate and strengthen its governance frameworks. This introspection is essential not only to uphold the standards and requirements of research degrees but also to ensure compliance with wider human rights, societal and ethical mandates. A notable example of national efforts to regulate generative AI comes from China, where the 2023 *Interim measures for the management of generative AI services* address key societal and ethical challenges, such as privacy protection, bias prevention, and the need for human oversight to ensure AI-generated content is accurate. These regulations emphasise transparency and accountability, preventing disinformation and protecting individual rights.

The American approach to generative AI policy is characterised by a mix of federal and state-level initiatives, reflecting both centralised and decentralised governance strategies. This is important to note as the separation of powers divides governance across executive, legislative and judicial branches as distinct departments of American government, establishing a system of 'checks and balances' to prevent one branch from becoming predominant. This essentially has resulted in broad frameworks as established by federal government (e.g., executive order 14105 from the Biden administration, 2023) in parallel with state governments who have been proactive in crafting their own AI regulations, often focusing on specific applications or sectors (e.g., California's Artificial Intelligence Accountability Act SB896, 2024). This dual approach allows for better tailoring of policies in accordance to local demands, but also underscores the growing complexity of varying state laws. Similarly, The United States of America (U.S) wider policy '*Blueprint for an AI Bill of Rights: Making automated systems work for the American people*' (2022) highlights the importance of privacy, algorithmic fairness, and protection from harmful generative AI systems. In the context of higher education, this framework advocates for the responsible use of generative AI, ensuring transparency and non-discrimination while mitigating risks of misinformation and bias in academic research. Together, these policies underscore global efforts to align generative AI use with human rights and societal ethics.

By proactively incorporating some of these issues, institutions have been able to better align their practices with evolving academic and ethical expectations, ultimately safeguarding the integrity and quality of their educational offerings and academic outputs. However, much of technological innovation has outpaced regulatory developments, highlighting the need for institutions to remain agile in times of complexity and uncertainty.

Within this landscape of change and ongoing policy review, there has surfaced an increased concern about digital inequity and access issues (Farrelly & Baker, 2023) for already marginalised student groups. The authors explain that researchers in well-funded institutions often have better access to advanced generative AI tools and high-performance computing resources, while those in underfunded or remote areas may struggle with limited facilities. The civic duty of universities to positively shape society and reduce inequalities is undeniable: there have been efforts across the sector to address digital inequities although in the UK, much of these efforts have been fragmented (Policy Connect, 2024). Regions such as sub-Saharan Africa, Central and South Asia are currently facing structural limitations such as lack of necessary infrastructure, data capabilities and skilled workforces needed for effective AI deployment (Yu & Qi, 2024). To bridge this gap, investments in local education and training opportunities is vital – universities can play a significant role in delivering on such initiatives, as evident by the explosion of Massive Open Online Courses (MOOCs) on the topic of AI. For example, the University of Helsinki opened their course, "Elements of AI" in 2018, which has been successful in targeting traditionally underrepresented groups in the field of science and technology. This includes female students (making up 37% of the course) and those over 45 (making up 21% of the course).

Similarly, the Commonwealth Artificial Intelligence Consortium was formed in 2023 to gather stakeholders from across global tech firms, research institutions and non-profits. Since its formation the Commonwealth secretariat has had an acute focus on delivering 'AI driven initiatives', such as the partnership with the University of West Indies and the SAS institute to grant access to data analytics tools and Microsoft Azure to students across the Caribbean. Such projects demonstrate the body's commitment to ensuring that digital skills gaps and technological inequalities are addressed.

Cloud-based solutions can help mitigate some access disparities by providing remote access to powerful generative AI tools, though reliable internet access remains a challenge. In a recent report published by JISC (2024) students mirrored these concerns regarding the increase in digital inequity that is slowly emerging, where those who have the financial means to pay for superior tools are gaining an academic advantage. Furthermore, a report by Policy Connect and The Higher Education Commission (2024) highlights the growing issue of digital poverty in UK higher education. It emphasises that digital poverty adversely affects students' access to online education, engagement with course materials and academic outcomes. Additionally, the report draws attention to the lack of access that impedes the development of essential ICT skills and digital literacy capabilities which are vital for levelling up a future jobs market. Findings from the report reflect current concerns published by the Office for Students (OfS), which outlines students are considered to be in digital poverty if they lack access to any of the following core components of digital infrastructure, including, the access to appropriate hardware, reliable internet connection, technical and repair services, a trained tutor and appropriate study spaces.

In response, the UK's Digital Development Strategy (DDS) for 2024–2030 focuses on digital inclusion to ensure no one is left behind, which aims to deliver of points outlined by the OfS report. This strategy emphasises affordable connectivity, digital literacy training, and support for underserved communities with the aim to reduce digital divides significantly by 2030 (UK Government, 2024a, 2024b). The UK's strategy for equitable access is comprehensive and ambitious, but its success hinges on additional funding and collaborative efforts within and between universities and industry partners, backed by government support, to address disparities and ensure all students have the resources needed to thrive in the digital age.

3 Prejudice and the Pursuit of Accurate Research Outcomes

One pressing issue is the potential for AI-generated outputs to perpetuate or even exacerbate existing biases present in the training data, leading to distorted research outcomes and reinforcing societal inequities that impact opportunities for postgraduate researchers (Rettberg et al., 2024) and their future careers. In the case study below, the author describes her role as a Senior Lecturer, working with postgraduate students in The School of Education, at The University of Greater Manchester. The author draws attention to enhancing students critical thinking using generative AI as a tool to address bias and representation in DALLE generative AI image creators. The author emphasises to postgraduate research students how these biases originate from AI's training data, which often underrepresent women and minorities or depicts them stereotypically. Consequently, generative AI tools used in academia may favour white, Western, male-authored papers in citations, skew grant allocations toward those same male groups, and limit mentorship opportunities for females. Similarly, racial biases lead to undervaluing work by minority researchers (Sun, L. et al., 2024). AI models, predominantly trained on English and Western-centric data, may fail to support non-Western scholars adequately. These systems can also reflect biases in academic metrics and peer review processes, further disadvantaging minority groups.

Postgraduate students are therefore in a niche position to drive the necessary changes within bias and distorted generative AI outputs. One critical approach is to be vigilant about the types of search results used when conducting research. For example, when carrying out literature reviews or gathering data, it is crucial to scrutinise sources for potential prejudice and ensure a balanced representation of diverse voices. This includes being critical of stereotypes in google image searches and questioning the representation of culture, race and gender within visual data, such as those found on Google Images, DALLE or other image generators which can perpetuate biased views. In a study by Guldberg (2022) the author concluded that Google Images, Google Maps, and Google Earth offer valuable resources for developing students critical thinking. The author emphasises the importance of integrating generative AI technology into teaching methods to foster a more engaging and effective learning environment and to help deconstruct bias and deep-rooted visual representations of the global north and south. These tools can help students analyse visual information, understand geographical contexts, and engage in interactive learning experiences.

As discussed, postgraduate students can play a pivotal role in changing the landscape of bias within research and knowledge exchange. However, for students to address these disparities, there needs to be a robust set of accountability mechanisms to ensure the reporting of generative AI training data that is diverse and ethical, and where applicable, regular bias audits should be conducted. Without such measures, generative AI risks entrenching inequities instead of helping to drive a more equitable academic world. One recommendation is to promote open access and community-driven publishing, as well as alternative publication platforms (Vackova et al., 2023). For example, the use of online digital 'zine' production that has the potential to engage culturally diverse contributors and further open up opportunities within the growing monetisation of the academic publication spheres. Furthermore, enhancing funding for research infrastructure in underserved regions and supporting community-based initiatives that foster diversity and inclusion within academia are essential measures. Despite the potential of emergent digital tools to democratise knowledge, barriers such as paywalls and subscription fees to academic journals, often restrict access to cutting-edge research, disproportionately affecting those at institutions with fewer resources and 'know how'. Continuous investment is required which includes dedicated leadership roles within generative AI, inclusive policy-making practices which are guided by input from key stake-holders including the importance of staff and student voices.

4 Originality and Authorship in Generative AI Enhanced Post Graduate Research

Johnson-Eilola and Selber (2007) outline there is entrenched view on authorship and originality in academia where the pursuit of writing is often associated with a romanticised notion of a solitary genius working in isolation. Nearly two decades later, this notion continues to subtly shape how the originality of work is perpetuated and framed within academic assessment frameworks. For example, university assessment policies and guiding principles for both staff and students often establish a dichotomy between "AI-assisted work" and "original work," which may inadvertently preserve the conventional hierarchy that favours exclusively self-produced content. As a consequence, students might feel apprehensive about utilising generative AI tools like Chat GPT and Grammarly even for legitimate educational purposes, as they are increasingly being suspected of using generative AI to 'cheat'. Equally, those students who opt to incorporate transparent uses of generative AI into their assessments could face stigma and be unfairly viewed as less autonomous, critical and proficient. Luo (2024) highlights a significant observation regarding the framing of academic notions of 'originality' and 'authenticity' within university policies. Luo (2024) found that some of these policies prioritised preventing academic misconduct over enhancing student learning through the use of generative AI. For example, in certain academic circles, institutions are advising teachers to redesign assessments to minimise cheating, emphasising 'policing' rather than engaging students in a positive way with its benefits. These institutional policy frameworks inadvertently embed negative language that builds mistrust amongst student groups. As noted by Adler-Kassner et al. (2008), institutions must focus on better ways to prevent plagiarism that move away from reinforcing the assumption that students are inherently prone to plagiarise.

Set within these changing times, the empirical case study presented below emphasises a shift towards developing human intelligences and advancing critical thinking skills, cross-disciplinary real-world projects, and social and emotional competencies (Luckin, 2025) within a generative AI enhanced landscape. Rather than prioritising technology itself, these human-centric areas need to take precedence and offer a way for postgraduate students to drive change.

5 Tackling AI Bias in Post Graduate Research: A Case Study from the Centre for AI in Education, The University of Greater Manchester, England

5.1 Current Institutional Challenges

In 2023, the Centre for AI in Education was established with a remit that included the formation of a cross-disciplinary team. This team was tasked with co-designing guidance for academic staff and postgraduate students on the integration of generative AI into teaching, research, and administrative practices. It was imperative for the guidance to speak to the strategic aims of the institution which emphasised employability, research impact, practical education, and industry partnerships, through continued teaching excellence in disciplines such as health, medicine, education, engineering, and business. The tone, content, and dissemination strategy of the guiding principles for staff and post graduate research students had to be both accessible and inclusive, aligning with the institution's rich history and its commitment to a widening participation and inclusivity.

The emergence of Artificial Intelligence (AI) sparked significant concern among postgraduate students many of whom were also staff members at the university who were undertaking continuous professional development. Research students felt apprehensive about keeping pace with digital advancements and addressing their perceived skills and knowledge gaps, while also managing ongoing accessibility issues. Additionally, there was growing discourse that generative AI systems were driving the change and not the other way around. Concerns also surfaced about the potential decline of the teaching role, with fears that generative AI might fundamentally take over marking and moderation, leaving students to discuss pastoral and academic issues solely with chatbots and lifeless digital interfaces.

5.2 *Actions Taken to Address the Challenges and Concerns*

Recognising these challenges, an opportunity opened up to reframe some of the negative discourse surrounding generative AI within academic spheres. The guiding principles would serve as a platform to build a positive rhetoric and prioritise human intelligence and real-world connections over the generative AI technology itself. Inspired by Luckin's (2020) work on enhancing various forms of human intelligences in a generative AI enhanced world. This approach aimed to prepare postgraduate research students to use generative AI as a tool, enabling them to reflect, critically review, and work cross-disciplinarily using authentic research practices and data collection methods.

The challenge was to shift the focus and dominant discourse from viewing generative AI as an autonomous system with control to emphasising human expertise, judgment, creativity, critical thinking, and personal reflection. Employability remained at the core of the agenda, with vital skills essential for the future labour market. We aimed to promote generative AI as a supportive tool, to be utilised only when necessary. For postgraduate students, it was crucial to balance this technological assistance with maintaining the integrity and originality of their research.

Building on the work of Guldberg (2022), students were encouraged to experiment with generative AI on a practical level, using the image creator, DALLE. The aim was to develop critical thinking skills regarding bias, prejudice and stereotypes in digital image creators. A series of workshops for postgraduate students were delivered to promote the university-wide guiding principles for the ethical integration of Generative AI. Postgraduate research students were asked to use a generative

AI image creator of their choice to use a generative AI image creator of their choice and to design representations of classrooms that were situated in the global South and the global North. Students were prompted to analyse how the images represented different race, cultures and teaching environments. Students were encouraged to critically engage with questions pertaining to social and cultural biases and consider ways to disrupt taken for granted notions of stereostypes and prejudices. Furthermore, students were asked to question what these search results revealed about the politics and economics of bias in our society and to discuss the differences in those representations. The goal was to inspire the same critical thinking in their own research when considering the types of data and searches they draw upon to inform their wider understanding. How might they recognise and disrupt these biases and thereby contribute to creating change?

5.3 *The Outcomes and Impact*

The main aim of these activities was to position postgraduate students as knowledge brokers, extending their influence beyond academic circles to challenge and transform entrenched biases and assumptions in the approach to generative AI search prompts. By emphasising the development of human intelligence and criticality in conjunction with generative AI, the objective was to empower students to preserve essential scholarly skills and maintain ethical standards within an increasingly AI-driven educational landscape.

5.4 *Navigating the Current Challenges in Generative AI Policy Development*

The rapid integration of Artificial Intelligence (AI) into academic research has necessitated the development of robust governance policies to ensure ethical use, maintain academic integrity, and foster innovation. Various systematic reviews that seek to evaluate the role of generative AI in academic research demonstrate the ubiquity of its use in research as well as the range of domains in which AI solutions are applied to augment scholarly activity (Khalifa & Albadawy, 2024; Bond et al., 2024). Generative AI has, and will continue to, profoundly challenge how academics and students approach research endeavours. As many of these papers point out, despite the intensification of generative AI applications across all disciplines and the ever-growing number of literature in and around this area, governance policies have remained underdeveloped and inconsistent.

Governance policies in generative AI for academic research should be designed to address the multifaceted challenges posed by AI technologies, including issues of accountability, transparency, and ethical considerations. Throughout this chapter we are building further awareness in exploring what these terms may refer to and how they can relate to research activities. Whilst these principles can be applied on an individual level, governance structures must be in place to ensure consistency in an approach during this transformative period wherein scholarly paradigms are shifting where generative AI has the potential to facilitate not only operational transformations (i.e., assistance with literature review or synthesis; data management and analysis support) but also marks a "methodological pivot of consequence" (Butson & Spronken-Smith, 2024, p. 565). Simply put, this refers to the fact that the emergence of generative AI has led to changes in the ways we approach research methods, and how this may have significant consequences for the ways researchers interact with AI tools for scholarly work. For example, generative AI systems enable complex, large-scale, and interdisciplinary analysis of phenomena, it can truly augment the scientific work academics and students partake in, rather than being a tool for automating repetitive tasks. As this change in thinking is still in progress, many higher education institutions and their leaders have been hesitant to readily accept its widespread adoption. In early 2023, 8 out of twenty-four UK Russell Group universities declared the use of generative AI solutions as academic misconduct (Chan, 2023), which has since been reversed. Similarly in the US, universities have been revising their academic integrity policies to address the use of generative AI with varying levels of severity (Huang, 2023). Universities in Asia have also placed temporary bans on using generative AI for coursework such as the University of Hong Kong and RV University in Bengaluru, India (Leung & Niazi, 2023). The retractions of many such decisions show an elevated level of uncertainty during this period of technological evolution.

To navigate these changes, the governance of generative AI in academic research should involve a combination of institutional policies, national regulations, and international guidelines. The following cases are some examples of how UK institutions have approached the development of such guidelines.

The University of Glasgow has developed comprehensive guidelines for the use of generative AI in research, underscoring the potential of AI to enhance and transform academic work. However, the university also recognises the disruptive potential of generative AI and the necessity of a balanced approach that incorporates careful and critical analysis of AI-generated outputs (University of Glasgow, 2024). The primary principle guiding the use of generative AI at the university is that any application of generative AI tools must be accompanied by critical oversight by the user. For example, generative AI tools should always supplement, not replace, the leading author or researcher of any given project. The guidance published on the university's website under their strategy and policies tab also require researchers to be transparent about their use of generative AI tools. This includes acknowledging generative AI contributions in their research and ensuring that AI-generated content is not presented as the researcher's own work, which would constitute academic misconduct. Transparency is particularly relevant for those who wish to continue their pursuit of academic research: credibility and integrity are integral to publications and scientific work to maintain the value of educational institutions (International Centre for Academic Integrity, 2021).

The promotion of academic integrity is central to the approach of the UK's Quality Assurance Agency for Higher Education (QAA). For context, the QAA is an independent body in the United Kingdom responsible for safeguarding standards and enhancing the quality of higher education. They have emerged as a pivotal force in shaping guidelines for the use of generative AI within the higher education sector through their "Quality Compass" publication from February 2024 (QAA, 2024).

The agency underscores the necessity for higher education institutions to establish robust policies that ensure generative AI tools are used in a manner that upholds academic honesty. This includes advising universities to incorporate generative AI usage into their academic integrity frameworks, thereby preventing academic misconduct and ensuring that AI-generated content is properly acknowledged and evaluated. The QAA also focuses on supporting both students and staff in navigating the complexities introduced by generative AI technologies. They provide a range of resources aimed at enhancing understanding and promoting responsible use of generative AI tools, which includes educational materials and webinars that explore the implications of generative AI for teaching, learning, and assessment. By fostering an environment of informed engagement, the QAA helps institutions adapt to the evolving educational landscape while maintaining high standards of quality and integrity.

The most novel part of the QAA's approach is their encouragement of educators to review traditional assessment methods. In an age where essays and other coursework can be automated in minutes, is the current approach to assessing students' abilities fit for purpose? Their guidance encourages the development of authentic assessment practices that account for the capabilities of generative AI tools, ensuring that evaluations accurately reflect student learning and understanding. This approach not only safeguards academic standards but also has the potential to prepare students for real-world applications of AI technologies. Development of generative AI literacy will become a critical aspect of higher education, as its use becomes more prevalent in the workplace and therefore an essential part of digital society (Southworth et al., 2023).

These are localised examples of governance policies in the UK, or guidelines that have been established to help students and staff understand how to better engage with AI tools. In parallel, there have been significant efforts from global organisations to develop international frameworks to emphasise the importance of aligning ethical standards across the world to safeguard academic and scientific integrity. For example, United Nations Educational, Scientific and Cultural Organisation (UNESCO) adopted the Recommendation on the Ethics of Artificial Intelligence, marking the first global framework in this domain (UNESCO, 2024). This framework emphasises the alignment of AI policies with existing human rights and international agreements, advocating for the responsible use of AI to benefit societies and democracies while promoting inclusive economic growth and environmental preservation. Not long after the recommendations were published, UNESCO released their roadmap for using AI in educational settings and research (Miao & Holmes, 2023), which marks their broader effort to ensure that AI technologies enhance human capacities and protect human rights, thereby promoting effective human-machine collaboration for sustainable development.

6 Understanding Policy Discourse in Postgraduate Study

Understanding generative AI policies ensures that postgraduate students can effectively and ethically leverage these technologies in their work. Students are primary stakeholders in educational institutions and should not be seen as passive recipients or 'objects' of policy and practice. As such, their involvement and experience of developing policy designs and strategies for generative AI adoption in coursework and research plays a vital role in the successful implementation and acceptance of such technologies (Valerio, 2024). By being informed about these policies and relevant developments, students can navigate the ethical challenges posed by generative AI, thereby safeguarding the integrity of their research. Conversely, by involving students in policy discussions, institutions can better align their strategies with the needs and expectations of the student body, fostering a more inclusive educational atmosphere. Furthermore, generative AI policy discourse provides students with insights into the broader implications of generative AI on society and the economy, preparing students for an AI-enriched future (Schiff, 2022). As future researchers and professionals, students need to understand how generative AI technologies can impact various sectors, influence policy decisions, and shape societal norms. Engaging with generative AI policy discourse equips them with the critical thinking skills necessary to assess the benefits and risks of generative AI applications, promoting informed decision-making and innovation that aligns with ethical and sustainable practices.

More importantly, awareness of and engagement in generative AI policy development empowers postgraduate students to contribute to the ongoing dialogue about the role of generative AI in education and research. This has been a central message throughout the chapter. As active participants in this discourse, postgraduate students can advocate for policies that support equitable access to generative AI resources and address global inequalities, reflecting UNESCO's commitment to inclusive and sustainable development. This engagement not only enhances their academic and professional growth but also positions them as informed citizens capable of influencing the future direction of generative AI technologies in society.

UNESCO has been central in developing guidance on an international scale. From 2023 to 2024, they have published four seminal books or reports on the topic of AI and Education. Starting from the 2019 *Beijing Consensus on Artificial Intelligence and Education* (2019), UNESCO documents have detailed the need for government to develop AI literacy and more advanced AI competencies as the technology sector advances. However, as the authors alluded to in this chapter, there are numerous frameworks all shaped by the context of the institution, and the higher education especially has yet to accept a sector-wide understanding of AI applications, and the skills required to navigate and confidently use products and tools.

Globally, there is variation on how countries define literacy, skills and competency, leading UNESCO to develop an AI competency framework for students in 2024. In the framework's introduction, the authors underscore the need to develop a "harmonised approach to integrating AI-related teaching and learning content in school curricula" (UNESCO, 2024, p. 12). Students are strongly encouraged to refer to their body of work to develop a human-centric understanding of AI utilisation for educational purposes.

What does a 'human-centric' understanding mean? In UNESCO's publications, it largely refers to the values, beliefs and critical thinking skills which are actively and purposefully applied to evaluating whether AI is fit for purpose, whether its use is justified, how humans should interact with it, and what responsibilities do actors have to the creation of safe, inclusive and just AI societies (ibid., p. 22). UNESCO's AI competency framework for students include the human centred mindset as one of the key pillars of AI literacy, demonstrating the importance of critical engagement with emerging technologies. There is tremendous value in developing a common language across all disciplines for understanding the application of AI and the products it generates: widely speaking, however, much of the discourse surrounding the linguistics of AI defines 'human-centric' AI as one that adheres to and even prioritises, ethical principles and social values.

Using key UNESCO documents and wider literature on AI literacy, we have developed a set of key recommendations for research students who are looking for practical steps forward in utilising AI for their work.

7 Key Recommendations for Postgraduate Research Students

1. **Use generative AI Tools to Enhance, not Replace:** While generative AI can significantly enhance research processes, it should not replace human judgment, creativity, and interaction. Postgraduate Students should use generative AI to complement their research, embedding human centric principles in utilising AI solutions to ensure that critical thinking, ethical reasoning, and community engagement remain central to their research endeavours.
2. **Understand Institutional Policies:** Familiarity with institutional generative AI policies is crucial as these guidelines dictate the acceptable use of generative AI tools in academic settings. Many institutions are still developing these policies, so it is important to keep up with the changes. Understanding what is expected from the institution and professors ensures that students use generative AI ethically and in compliance with academic standards. This awareness helps prevent misuse and supports the integrity of any research outputs or coursework.
3. **Understand the Capabilities and Limitations of generative AI Tools:** Generative AI tools offer powerful capabilities, such as automating data analysis and literature reviews, but they also have limitations, including potential biases

and the "black box" problem where the decision-making process is not transparent (see Varsha, 2023). Students should be aware of these factors to use generative AI effectively and critically, ensuring that their research conclusions are valid and reliable – overreliance can be dangerous especially when trying to produce original work.

4. **Engage with Ethical Considerations:** Ethical use of generative AI involves understanding issues such as data privacy, consent, and bias. Students must consider these ethical dimensions to ensure their research practices do not harm individuals or communities and contribute positively to the body of knowledge. Being reflexive about your position as a researcher and the tools you engage with is essential in maintaining a level of academic integrity and awareness of the impact your research findings may have (for further guidance on reflexivity, see Jamieson et al., 2023).
5. **Seek Support and Training:** Engaging with tutors, professors, and library services, and seeking training on generative AI tools can enhance students' ability to use these technologies effectively that are especially tailored to the diverse needs of researchers in wide ranging disciplines (Chubb et al., 2022). Educators can provide guidance on best practices and help students navigate complex generative AI systems, ensuring that they are well-equipped to integrate generative AI into their research responsibly – that is what they are there for!
6. **Collaborate with Peers and Share Best Practice:** Postgraduate students are ahead of the curve in many ways, utilising generative AI tools in more sophisticated ways than expected. Collaboration with peers allows students to share insights and strategies for using generative AI tools effectively (for a case study, see Joseph et al., 2024). This exchange of knowledge can lead to innovative approaches and solutions, fostering a community of practice that enhances learning and research outcomes.

References

Adler-Kassner, L., Anson, C.M. and Howard, R.M. (2008). Framing plagiarism. Originality, imitation, and plagiarism: *Teaching writing in the digital age*, pp.231–246.

Butson, R., & Spronken-Smith, R. (2024). AI and its implications for research in higher education: A critical dialogue. *Higher Education Research & Development*. https://www.tandfonline.com. https://doi.org/10.1080/07294360.2023.2280200

Bond, M., Khosravi, H., De Laat, M., Bergdahl, N., Negrea, V., Oxley, E., Pham, P., Chong, S. W., & Siemens, G. (2024). A meta systematic review of artificial intelligence in higher education: A call for increased ethics, collaboration, and rigour. *International Journal of Educational Technology in Higher Education, 21*(1), 4. https://doi.org/10.1186/s41239-023-00436-z

Burger, B., Kanbach, D. K., Kraus, S., Breier, M., & Corvello, V. (2023). On the use of AI-based tools like ChatGPT to support management research. *European Journal of Innovation Management, 26*(7), 233–241. https://doi.org/10.1108/EJIM-02-2023-0156

Capraro, V., Lentsch, A., Acemoglu, D., Akgun, S., Akhmedova, A., Bilancini, E., Bonnefon, J.-F., Brañas-Garza, P., Butera, L., Douglas, K. M., Everett, J. A. C., Gigerenzer, G., Greenhow, C., Hashimoto, D. A., Holt-Lunstad, J., Jetten, J., Johnson, S., Kunz, W. H., Longoni, C., et al. (2024). The impact of generative artificial intelligence on socioeconomic inequalities and policy making. *PNAS Nexus, 3*(6), pgae191. https://doi.org/10.1093/pnasnexus/pgae191

Chan, C.K.Y. (2023). Is AI changing the rules of academic misconduct? An in-depth look at students' perceptions of 'AI-giarism'. arXiv:2306.03358. Department for Digital, Culture, Media & Sport. (2023). 'Digital development strategy 2024 to 2030'. UK Government. https://www.gov.uk/government/publications/digital-development-strategy-2024-to-2030/digital-development-strategy-2024-to-2030.

Chen, Y., Kirshner, S., Ovchinnikov, A., Andiappan, M. and Jenkin, T. (2024). Biases of Humans, of AI, and of Humans with AI. Available at SSRN 4943377.

Chubb, J., Cowling, P. and Reed, D. (2022). Speeding up to keep up: exploring the use of AI in the research process. *AI & society, 37*(4), pp.1439–1457.

Dwivedi, Y. K., Hughes, L., Ismagilova, E., Aarts, G., Coombs, C., Crick, T., et al. (2021). Artificial Intelligence (AI): Multidisciplinary perspectives on emerging challenges, opportunities, and agenda for research, practice and policy. *International Journal of Information Management, 57*, 101994.

Farrelly, T., & Baker, N. (2023). Generative artificial intelligence: Implications and considerations for higher education practice. *Education Sciences, 13*(11), 1109.

UK Government. (2024a). Digital development strategy 2024-2030. https://www.gov.uk/government/publications/digital-development-strategy-2024-to-2030/digital-development-strategy-2024-to-2030.

Guldberg, C. (2022). Using Google images, maps and earth to teach critical thinking: Decolonising the curriculum and beyond. *Journal of PGR Pedagogic Practice, 2*, 61–78.

Huang, Kalley. (2023, Jan 19) "Alarmed by A.I. Chatbots, universities start revamping how they teach.", *International New York Times.* Retrieved from https://www.nytimes.com/2023/01/16/technology/chatgpt-artificial-intelligence-universities.html. Accessed August 15, 2024.

International Center for Academic Integrity [ICAI]. (2021). *The fundamental values of academic integrity.* (3rd ed.). Retrieved from: http://www.academicintegrity.org/the-fundamental-values-of-academic-integrity/.

Jamieson, M. K., Govaart, G. H., & Pownall, M. (2023). Reflexivity in quantitative research: A rationale and beginner's guide. *Social and Personality Psychology Compass, 17*(4), e12735. https://doi.org/10.1111/spc3.12735

Jisc. (2024, May). Student perceptions on generative AI. Available at: https://www.jisc.ac.uk/report/student-perceptions-on-generative-ai. Accessed June 3, 2024.

Johnson-Eilola, J., & Selber, S. A. (2007). Plagiarism, originality, assemblage. *Computers and Composition, 24*(4), 375–403.

Joseph, G. V., Athira, P., Anit Thomas, M., Jose, D., Roy, T. V., & Prasad, M. P. (2024). Impact of digital literacy, use of AI tools and peer collaboration on AI assisted learning-perceptions of the university students. *Digital Education Review, 45*, Article 45. https://doi.org/10.1344/der.2024.45.43-49

Juarez-Orozco, L. E., Martinez-Manzanera, O., Storti, A. E., & Knuuti, J. (2019). Machine learning in the evaluation of myocardial ischemia through nuclear cardiology. *Current Cardiovascular Imaging Reports.* https://doi.org/10.1007/s12410-019-9480-x

Khalifa, M., & Albadawy, M. (2024). Using artificial intelligence in academic writing and research: An essential productivity tool. *Computer Methods and Programs in Biomedicine Update, 5*, 100145. https://doi.org/10.1016/j.cmpbup.2024.100145

Kox, E.S. & Beretta, B. (2024). Evaluating Generative AI Incidents: An Exploratory Vignette Study on the Role of Trust, Attitude and AI Literacy. In HHAI 2024: Hybrid Human AI Systems for the Social Good (pp. 188–198). IOS Press.

Liu, J., & Jagadish, H. V. (2024). Institutional efforts to help academic researchers implement generative AI in research. *Harvard Data Science Review*, Special Issue 5. https://doi.org/10.1162/99608f92.2c8e7e81

Luckin, R. (2020). Developing human intelligence in an AI intelligent world. In *Human learning in the age of artificial intelligence* (pp. 95–117). Springer.

Luckin, R. (2025). Nurturing human intelligence in the age of AI: rethinking education for the future. Development and Learning in Organizations: *An International Journal, 39*(1), pp.1–4.

Luo, J. (2024). A critical review of GenAI policies in higher education assessment: A call to reconsider the "originality" of students' work. *Assessment & Evaluation in Higher Education*, 1–14.

Leung, M., & Niazi, S. (2023, February 24). Universities on alert over ChatGPT and other AI-assisted tools. University World News. Retrieved from https://www.universityworldnews.com/post.php?story=20230222132357841.

Memarian, B., & Doleck, T. (2023). Fairness, accountability, transparency, and ethics (FATE) in artificial intelligence (AI) and higher education: A systematic review. *Computers and Education: Artificial Intelligence, 5*, 100152. https://doi.org/10.1016/j.caeai.2023.100152

Miao, F., & Holmes, W. (2023). *Guidance for generative AI in education and research*. UNESCO.

Müller, H., Pachnanda, S., Pahl, F. B., & Rosenqvist, C. (2022). The application of artificial intelligence on different types of literature reviews—a comparative study. *International Conference on Applied Artificial Intelligence, 2022*, 1–7.

Nah, F., Zheng, R., Cai, J., Siau, K., & Chen, L. (2023). Generative AI and ChatGPT: Applications, challenges, and AI-human collaboration. *Journal of Information Technology Case and Application Research, 25*(3), 277–304. https://doi.org/10.1080/15228053.2023.2233814

Neyedli, H. F., Hollands, J. G., & Jamieson, G. A. (2011). Beyond identity: Incorporating system reliability information into an automated combat identification system. *Human Factors, 53*(4), 338–355. https://doi.org/10.1177/0018720811413767

Nguyen-Trung, K., Saeri, A. K., & Kaufman, S. (2023). Applying ChatGPT and AI-powered tools to accelerate evidence reviews. https://doi.org/10.31219/osf.io/pcrqf.

Özer, M., Perc, M. & Suna, H.E. (2024). Artificial intelligence bias and the amplification of inequalities in the labor market. *Journal of Economy Culture and Society, (69)*, pp.159–168.

Pal, S. (2023). A paradigm shift in research: Exploring the intersection of artificial intelligence and research methodology. *International Journal of Innovative Research in Engineering & Multidisciplinary Physical Sciences, 11*(3). https://doi.org/10.37082/IJIRMPS.v11.i3.230125

Policy Connect, & Higher Education Commission. (2024). Digitally enhanced blended learning: Leveraging the benefits of technology in higher education. Policy Connect. Retrieved from https://www.policyconnect.org.uk/events/report-launch-digitally-enhanced-blended-learning.

QAA (2024a). Quality Compass: Navigating the Complexities of the AI Era in Higher Education. Gloucester: Quality Assurance Agency for Higher Education. Available at: https://www.qaa.ac.uk/news-events/news/new-qualitycompass-on-artificial-intelligence.

Rettberg, J. W., Crawford, K., Schultz, J., Taylor, L., Suchman, L., Andrejevic, M., & Ananny, M. (2024). An AI Society: Artificial intelligence is reshaping society, but human forces shape AI. Social scientists and humanities experts explore how to harness the interaction, revealing urgent avenues for research and policy. *Issues in Science & Technology, 40*(2).

Schiff, D. (2022). Education for AI, not AI for education: The role of education and ethics in national AI policy strategies. *International Journal of Artificial Intelligence in Education, 32*(3), pp. 527–563.

United Nations Educational, Scientific and Cultural Organization [UNESCO]. (2024). AI competency framework for students, : France, https://doi.org/10.54675/JKJB9835.

University of Glasgow. (2024, April 15). Generative AI Guidance for Researchers. Research Services Directorate, Retrieved from https://www.gla.ac.uk/research/strategy/ourpolicies/ai-for-researchers/#definitions%2Fglossaryofusefulterms.

Selwyn, N. (2022). The future of AI and education: Some cautionary notes. *European Journal of Education, 57*(4), 620–631. https://doi.org/10.1111/ejed.12532

Smith, S., Tate, M., Freeman, K., Walsh, A., Ballsun-Stanton, B., Hooper, M., & Lane, M. (2024). A university framework for the responsible use of generative AI in research. *arXiv Preprint*, arXiv:2404.19244.

Southworth, J., Migliaccio, K., Glover, J., Glover, J., Reed, D., McCarty, C., Brendemuhl, J., & Thomas, A. (2023). Developing a model for AI Across the curriculum: Transforming the higher education landscape via innovation in AI literacy. *Computers and Education: Artificial Intelligence, 4*, 100127. https://doi.org/10.1016/j.caeai.2023.100127

Spiegelhalter, D. (2020). Should we trust algorithms? *Harvard Data Science Review, 2*(1). https://doi.org/10.1162/99608f92.cb91a35a

Sun, L., Wei, M., Sun, Y., Suh, Y. J., Shen, L., & Yang, S. (2024). Smiling women pitching down: auditing representational and presentational gender biases in image-generative AI. *Journal of Computer-Mediated Communication, 29*(1), zmad045.

UK Government. (2024b). *Digital development strategy 2024–2030.* Foreign, Commonwealth & Development Office. https://assets.publishing.service.gov.uk/media/6613e7f7c4c84d4b31346a68/FCDO-Digital-Development-Strategy-2024-2030.pdf.

UNESCO, 2024. AI Competency Framework for Students. Paris: United Nations Educational, Scientific and Cultural Organization.

Vackova, P., Puntil, D., Dowdeswell, E., Cooke, C., & Caton, L. (2023). Collaborative writing as bio-digital quilting: A relational, feminist practice towards "Academia Otherwise". *Social Inclusion, 11*(3), 65–76.

Valerio, A. (2024). Anticipating the impact of artificial intelligence in higher education: student awareness and ethical concerns in zamboanga city, philippines. *Cognizance Journal of Multidisciplinary Studies, 4*(6), 408–418. https://doi.org/10.47760/cognizance.2024.v04i06.024

Varsha, P. S. (2023). How can we manage biases in artificial intelligence systems–A systematic literature review. *International Journal of Information Management Data Insights, 3*(1), p.100165.

Yu, J. & Qi, C. (2024). The impact of generative AI on employment and labor productivity. *Review of Business, 44*(1), pp. 53–67.

Chapter 9
Data Integrity and AI Ethics

Khaldoon Dhou and Leyla Orudzheva

Abstract This chapter addresses the vital importance of data in AI research, focusing on data collection methods, integrity, and the ethical considerations involved, especially concerning privacy and bias in AI algorithms. It intends to heighten awareness of the ethical implications of data usage in AI, with strategies for maintaining data integrity and reducing bias. A comprehensive discussion on ethical data practices will highlight their critical role in conducting responsible AI research.

1 Introduction

"...data is the new oil and artificial intelligence is the refinery..." (Jones, 2023). The recent advancements in GenAI have significantly impacted how AI is employed across various industries and disciplines. GenAI tools have become highly valuable for scientists, allowing them to navigate their research projects efficiently and precisely. As AI continues to evolve, it is crucial to recognize its potential benefits and ethical challenges, especially regarding data integrity and bias mitigation. Any research design should strive to pay attention to data, including how it is sampled and how, what, and why it is being collected, to ensure both internal and external validity as well as construct validity (Bhattacherjee, 2012). The widespread use of AI tools for research data collection, processing, and predictions may seem like a magic solution that can make powerful predictions using the myriads of data available through the Internet of Things (IoT), social media inputs, and other data points that are constantly accumulated by businesses, governments, and academics alike. Yet, without ensuring that the data is right for a given problem and the basis of predictions is solid so as not to contradict the existing body of scientific knowledge, the predictions generated by AI tools may be misleading as the decisions based on those

K. Dhou (✉) · L. Orudzheva
Texas A&M University Central Texas, Killeen, TX, USA
e-mail: kdhou@tamuct.edu

X. Zhou, H. Al-Samarraie (eds.), *Institutional guide to using AI for research*, Advances in Artificial Intelligence in Education 2,
https://doi.org/10.1007/978-3-031-94809-1_9

predictions (Redman & Hoerl, 2024). With the increasing reliance on GenAI, researchers must navigate a complex landscape where data quality and ethical considerations are paramount (Kirova et al., 2023; Poth et al., 2024). It is essential that AI-driven technologies maintain accuracy, fairness, and transparency to sustain responsible practices in the field (Walmsley, 2021).

In the following chapter, we will review the fundamental data governance framework, data collection methods, and types of data commonly used in AI research. Next, we will probe legal and ethical considerations related to AI data collection and use, focusing on issues of copyright ownership, transparency, and informed consent. We will continue with a discussion of data integrity with an emphasis on data validation and security and will review key aspects of model interpretation. We conclude with a recapitulation of ethical guidelines for data in AI research and provide examples of several case studies drawn from multiple disciplines to illustrate how data is used and whether best practices are being followed.

2 Data Governance Framework

Data governance serves as a crucial framework that delineates the authority and responsibility for decision-making concerning an organization's data (Ahmadi et al., 2022). By setting forth clear rules and guidelines, it ensures that the right individuals are accountable for data-related decisions. As data becomes increasingly vital across various sectors, the significance of solid data governance has escalated. This framework extends beyond conventional data management, emerging as a vital element in fields such as data science, sustainable development (Singhal et al., 2024), and Internet of Things (IoT) computing (Pansara, 2023), where the management of extensive and intricate data sets is pivotal to achieving targeted outcomes. In AI, data governance assumes a heightened level of importance. AI systems heavily depend on vast datasets to produce accurate and meaningful results, making the assurance of data quality and integrity essential (George et al., 2017). Within this context, data governance is a foundational framework that directs ethical data collection and management practices, which are critical for developing reliable AI systems. Without a robust governance structure, organizations risk subjecting their AI systems to biases, inaccuracies, and potential breaches, which could compromise the effectiveness of AI applications. Thus, data governance goes beyond being a mere set of guidelines; it functions as a safeguard that preserves the reliability of AI outcomes, thereby sustaining trust among users and the public.

The significance of data governance becomes even more evident in fields such as healthcare and finance, where data quality is crucial for the effectiveness and fairness of AI models. For example, AI systems that rely on biased or incomplete data in healthcare may result in incorrect diagnoses or treatment recommendations, potentially endangering patient well-being. Similarly, where transparency and accountability are paramount in the finance sector, data governance ensures that AI-driven decisions adhere to robust ethical standards (Jauhiainen & Lehner, 2022).

By upholding stringent ethical practices, organizations can address issues related to poor data quality or mismanagement, ensuring that AI remains a constructive tool rather than a source of new ethical challenges. To tackle these challenges, organizations must focus on the quality of data and the methods used for its collection and management (Janssen et al., 2020). Effective data governance involves implementing comprehensive strategies to address specific data management issues (Abraham et al., 2019). These strategies often encompass formal structures, well-defined processes, and best practices that promote effective stakeholder collaboration. Additionally, a robust data governance framework should include a system of rewards and penalties to encourage responsible data handling (Janssen et al., 2020). By adopting such measures, organizations can build a culture of accountability and continuous improvement in data practices, which is crucial for maintaining the integrity of AI systems. Beyond the boundaries of individual organizations, data governance also encompasses a broader network of interconnected entities. In the current digital environment, effective data governance necessitates collaboration among various organizations and individuals, establishing a network of trust and accountability (Janssen et al., 2020). Creating trusted frameworks for secure and reliable data exchange is crucial in this regard. These frameworks ensure that data is shared responsibly and that all parties adhere to the General Data Protection Regulation (GDPR) and other pertinent legal standards. By following these frameworks, organizations can promote secure and ethical data-sharing practices, vital for sustaining trust and cooperation across different sectors.

3 Data Collection Methods

Researchers employ various methods to gather data for their studies, each suited to different needs and contexts. One approach involves traditional techniques where researchers collect specific types of data for well-defined objectives. For example, an organization might use surveys and interviews or compile structured data categorized into predefined formats. This method usually results in structured data, such as that found in databases or spreadsheets. While practical and still prevalent, this approach often limits the volume of records that can be collected. On the other hand, AI-driven methods facilitate the collection of large-scale data from diverse sources, including websites and computer simulations. AI systems excel in processing vast amounts of data in real-time, making them particularly useful for scenarios where data is generated rapidly, such as on social media platforms. Extensive research reveals that AI data collection techniques can be grouped into several categories: passive data collection, synthetic and simulated data, public and open data, crowdsourced and user-generated data, and modeling methods. Below is an overview of each category:

3.1 Passive Data Collection

This method, often referred to as indirect data collection, involves gathering data without direct input from participants. Typically, this data is collected automatically as users interact with various systems. Researchers utilize a range of techniques for passive data collection, such as web scraping, sensor data collection, log data analysis, and behavioral tracking. While passive data collection methods can be advantageous in terms of scale and scope, they present ethical challenges. That is to say, although passive data collection does not involve active participation, it is essential to maintain transparency (Maher et al., 2019). In other words, users need to be aware that their data is being collected. This can be done by offering clear privacy policies and notifications that explain to users how their data is being collected and how it is used. It is crucial to provide users with this information to increase their trust in an organization (Morey et al., 2015).

3.2 Synthetic and Simulated Data

Synthetic and simulated data (also called silicon samples) involves generating data through AI techniques as a substitute for real-world data. This approach is particularly valuable when collecting actual data is impractical for addressing certain research questions. GenAI has become one of the primary tools for generating synthetic data, as it employs advanced machine learning models to create new data points that mimic the statistical properties of real-world datasets (Voola et al., 2024). GenAI is particularly useful when large amounts of diverse data are needed for training machine learning models, without relying on sensitive or difficult-to-collect real data (Lee, 2023). A notable example is in the domain of chess, a field extensively studied by both computer scientists and psychologists. Research has examined various strategies employed by chess players and their impact on opponents of varying skill levels. For instance, some players might adopt an aggressive approach. In contrast, others may lean towards a defensive strategy (Caci & Dhou, 2020; Dhou, 2018b, 2020b, 2021a, 2021b, 2023). AI offers significant flexibility in designing experiments that simulate encounters between players from different eras. Virtual chess simulations can recreate the play styles of grandmasters such as Kasparov, allowing researchers to explore chess strategies and outcomes in ways that might be challenging with real-world data (Dhou, 2019). This research can yield valuable insights for business and other fields interested in understanding strategic behaviors. While GenAI-driven synthetic data offers a great deal of flexibility, bias is a major concern when generating synthetic data, as algorithms might introduce biases within the model. AI tools have been shown to contain embedded biases related to political perspectives (Marcus, 2024) and cultural norms that tend to be highly influenced by American and some European cultures (Durmus et al., 2023). Data labeling and algorithms created for law enforcement purposes, such as those based on historical policing records or facial recognition, display skin type and gender biases that are problematic (DoCarmo et al., 2021). This poses significant ethical concerns with regard to the data originated from or processed by such tools and raises the issue of fairness (Maslej et al., 2024). To address this issue, researchers can consider employing probabilistic methods to identify data samples that are hard to predict and then applying boosting (Draghi et al., 2024). Researchers should actively implement rigorous checks for bias while data is being generated. In the case of GenAI, regular auditing of the model's output and employing bias-detection algorithms is likely to mitigate some of these concerns (Ferrara, 2023; Huang et al., 2024). Additionally, while synthetic and simulated data can be robust in making decisions, it is essential to mention that their limitations need to be understood. In other words, although they can provide valuable insights, decision-makers should be careful when using them in specific domains, such as healthcare and criminal justice.

3.3 Public and Open Data

Public and open data refers to datasets that are freely accessible and typically provided by governments or organizations. These datasets are available for researchers to use in various AI applications. Although this type of data can be considered the foundation for many types of projects, it has some issues, such as inconsistent formats that cause issues when incorporated within AI models (Jiang et al., 2021). These public datasets might have biases that come from data collection, which can eventually cause biased outcomes.

3.4 Crowdsourced and User-Generated Data

Crowdsourced and user-generated data typically involves contributions from a diverse group of individuals, either through crowdsourcing efforts or by sharing content on digital platforms. This type of data can be used for a wide range of purposes. For example, Sosko and Dalyot (2017) explored how weather data gathered from mobile device sensors, contributed by users through crowdsourcing, can be used to develop a comprehensive and dense sensor network. The employment of crowdsourced and user-generated data offers some benefits. However, it brings its own challenges, especially in terms of the accuracy and reliability of data. One example is urban flood prediction, where fine-grained hydrological data are essential but often unavailable via traditional methods. To fill out these gaps, crowdsourcing from webcams, social media platforms, and others is an option (Helmrich et al., 2021). On the other hand, the consistency and reliability of this data create difficulties in the process. As shown in research, incorporating multiple streams can overcome some of these issues (Helmrich et al., 2021). However, it is crucial to address data validation and standardisation to ensure reliable results.

Bias in crowdsourced data adds further complications, as the data itself can be shaped without the intention of the creators. For example, during natural disasters, social media activity tends to concentrate on sharing information about risks and helping the community cope with risks. Research shows that despite the spread of social media, areas with lower populations or low-income areas receive less attention, which can skew the data and negatively affect the response efforts (Fan et al., 2020). To address these issues, Fan et al. (2020) examined the attention given to locations based on social media content and considered content-based analysis to filter tweets that belong to different neighbourhoods. The study is valuable in the sense that it provides insights into the distribution of social media activities in neighbourhoods affected by disasters.

3.5 Agent-Based Modeling Approaches

A new perspective of using AI methods for data collection is through agent-based modeling approaches. An agent-based model is a simulation technique that explores the interactions between various entities, such as people, animals, and environments. A prominent tool in this domain is NetLogo, a platform equipped with preloaded models across fields like computer science, biology, and psychology. In NetLogo, researchers can run simulations with chosen parameters to investigate specific phenomena. For example, one biological simulation in NetLogo models an ant colony with multiple food sources. This simulation illustrates how ants release pheromones to guide them to new food sources. Inspired by this, a new system was developed that uses the concept of pheromones and ant movements to collect data from bi-level images (Mouring et al., 2018). Beyond simulating ant colonies, NetLogo has inspired the development of various techniques for collecting data from digital imaging. One notable approach is the use of predator-prey ecosystems to encode image information (Dhou, 2018a, 2020a). This technique involves converting an image into a virtual environment populated by static agents, such as sheep and paths, and introducing wolves as dynamic agents tasked with hunting the sheep. The algorithm tracks the wolves' movements and processes and compresses the data, potentially surpassing established benchmarks in bi-level compression. Similarly, Dhou and Cruzen (2020) introduced a compression system based on dolphin echolocation. This system transforms a bi-level image into a virtual setting where dolphins use sound waves to detect nearby fish while moving. The approach demonstrated success in encoding image information and achieving high compression rates. Other methods for data collection that are inspired by biological behaviors include simulations of bats (Dhou & Cruzen, 2022c), solitary behavior of tigers (Dhou & Cruzen, 2022b), HIV modeling (Dhou & Cruzen, 2021b, 2022a), beaver territories (Dhou & Cruzen, 2021a), and natural reproduction (Dhou, 2019). It is essential to mention that ethical issues need to receive serious attention when employing agent-based modeling for data collection in fields such as healthcare and finance. This is because agent-based modeling simulates the behavior of complex entities within controlled environments, and making assumptions might introduce biases to the results. For example, when human behavior is being simulated based on biased data, the results can reinforce these biases. These issues are particularly concerning when the outcomes are to be used by policymakers and can significantly impact healthcare decisions. In Table 9.1, we provide an overview of various data collection methods, highlighting their key characteristics, applications, advantages, and challenges. This comparison aims to help researchers understand the strengths and limitations of each approach, allowing for more informed decisions when selecting appropriate data collection techniques for their studies.

Table 9.1 Comparison of data collection methods

Method	Description	Applications	Advantages	Challenges
Passive data collection	Data gathered automatically without direct user involvement, e.g., sensors, web scraping	Behavioral research, real-time monitoring	Unobtrusive, scalable	Requires transparency, consent, and careful privacy considerations
Synthetic and simulated data	AI-generated data that mimics real-world data, often used when collecting real data is impractical	Training AI models, testing algorithms, simulations (e.g., chess strategies)	Scalable, can address data scarcity	Risk of bias, may not fully capture real-world nuances
Generative AI (GenAI)	Advanced AI models that generate synthetic data to reflect real-world patterns	Healthcare simulations, creating training data for machine learning	Flexible, reduces reliance on sensitive or hard-to-collect real data	Potential bias, especially in output data if not audited regularly
Public and open data	Freely available datasets, often provided by governments or institutions	Academic research, public policy analysis	Easily accessible, large datasets	Inconsistent formats, data may be biased based on collection methods
Crowdsourced and user-generated data	Data contributed by users through crowdsourcing or content shared on platforms	Market research, urban planning, social science	Diverse data sources, low-cost	Data accuracy and reliability can vary, validation is needed
Agent-based modeling	Simulation of interactions between autonomous agents to explore complex systems	Biological simulations, economic models, and digital imaging analysis	Allows detailed, dynamic analysis of complex behaviors	Ethical concerns in sensitive fields (e.g., healthcare), may introduce biases in assumptions

4 Legal and Ethical Considerations

A major area of concern originates in the manner AI tools use existing data content to train their algorithms as they usually access such data from publicly available sources without explicit consent of participants or copyright owners or any license for using material for AI training purposes (Islam & Greenwood, 2024; Maslej et al., 2024). For example, Nvidia scrapped YouTube downloading all the video content it contains to train its video generating AI tool (Gilbert, 2024). Similarly, companies like Meta, Apple and Google tap into a wealth of user data to train their AI tools without clearly explaining what and how is being used nor making it easy for users to opt out for sharing their data (Fung, 2024; Grant, 2024; Jiménez, 2024). Even though such practices may seem concerning for a portion of their respective users, even more concerning is the AI tools that may result out of them as well as the data that may spill over into scientific research without proper checks and balances. Moreover, "self-learning capacities of algorithms and the autonomy of AI systems that results from these capacities" make the process more opaque and harder to replicate (Buhmann & Fieseler, 2023). Importantly, the quality of data inputs has crucial implications for the consecutive inferences AI makes out of the data, sometimes assuming causality due to strong correlations while a poorly designed AI can even lead to harmful outcomes (Buhmann & Fieseler, 2023). While academic internal review boards typically oversee primary data collection ensuring that a proper informed consent has been obtained and the study participants are not being harmed, use of secondary data obtained from 3rd party sources or scrapped directly in the web, whether for primary analysis or for AI tool training purposes, add to data transparency concerns (DoCarmo et al., 2021).

This has further implications for the legality of the data obtained as multiple jurisdictions implement laws putting constraints on user data collection and storage, e.g., General Data Protection Regulation (GDPR), California Consumer Privacy Act (CCPA), with certain data collection procedures and assessment tools ruled unconstitutional in the United States.

(DoCarmo et al., 2021) while others require generative ai providers disclose what copyrighted data was used during their tool training (Thongmeensuk, 2024). However, some jurisdictions have provisions for text and data mining (TDM) exemptions either for research purposes, non-commercial use or even use of work for nonenjoyment purpose and the principle of fair use allows to exploit copyrighted material under certain circumstances (Thongmeensuk, 2024). A comprehensive list of existing ethics guidelines for AI by country of issuer including private and government entities is compiled by Jobin and colleagues (Jobin et al., 2019).

5 Data Integrity

Data integrity encompasses the set of principles and practices that are crucial for maintaining the credibility, reliability, and ongoing usefulness of data within databases (Ugboma, 2004). Upholding data integrity ensures that information remains accurate, consistent, and trustworthy, which is vital for making well-informed decisions and fostering confidence in the systems that manage the data. Furthermore, it enables organizations and researchers to adhere to rigorous standards, safeguard against fraud, and ensure the long-term value of their data.

Maintaining data integrity involves essential practices to ensure the accuracy and reliability of information. This includes verification and validation processes. Verification focuses on ensuring that data is correctly entered and adheres to the required format (Maaliw III & Degadwala, 2023). Validation, on the other hand, involves confirming that the data used in constructing and evaluating models is both appropriate and accurate (Araújo et al., 2008). For instance, in the realm of Wireless Sensor Networks (WSNs), verifying and validating data is crucial to uphold the integrity of fragmented data blocks. During data fragmentation, there's a risk that some blocks might be lost or compromised by forged data injected by attackers. To mitigate this risk, Daniel D and Roslin (2021) developed a method to verify the integrity of each fragmented block and validate the data's accuracy. Their approach helps prevent unauthorized data alterations and ensures that only accurate data is used in aggregation. Their simulation results demonstrated improvements in data correctness and reliability.

In addition to verification and validation, data cleaning is crucial for maintaining data integrity. Data cleaning involves identifying and rectifying errors, inconsistencies, and redundancies in datasets. While violations of integrity constraints are a form of data error, other issues like missing values and duplicates also fall under this category (Rezig et al., 2019). Effective data cleaning ensures that datasets remain accurate and well-organized, thereby facilitating better data analysis and decision-making. Although many data cleaning processes involve human intervention, algorithms typically handle the bulk of the cleaning tasks, with humans providing final oversight to verify data accuracy (Rezig et al., 2019). Given the impracticality of manually correcting all data, many techniques incorporate human feedback to enhance the efficiency of the data cleaning process.

Data security is another key aspect of maintaining data integrity. Implementing strong security measures—such as access controls, encryption, and regular audits—is essential to safeguard data from unauthorized access. For instance, Lu et al. (2020) proposed a decentralized data integrity auditing scheme designed to enhance both the security and scalability of verifying data stored on remote servers. Their approach, which involves efficient auditing protocols, demonstrated significant improvements in both security and performance. The results highlighted that their method was robust in maintaining data integrity and proved effective in practical performance evaluations. Another significant threat to data security is the possibility to extract sensitive training data from the public-facing interface of an AI tool like

ChatGPT by using "attack" techniques provoking the model to behave in a way it was not initially meant to by the algorithm designers (Nasr et al., 2017).

In summary, upholding data integrity involves integrating verification and validation processes, data cleaning techniques, and robust data security measures. These practices are vital for ensuring that data remains accurate and reliable, safeguarding it against potential threats, and supporting managers in making well-informed decisions.

Ensuring data integrity poses considerable challenges for organizations. One key issue is keeping data accurate and consistent across distributed systems like Internet of Things (IoT) devices. Over time, data can become corrupted, incomplete, or inconsistent, which undermines the reliability of decisions made based on that data (Xie et al., 2016). Another major challenge is regulatory compliance; navigating the complex landscape of data protection and privacy laws can be overwhelming. Failing to comply with these regulations not only risks legal penalties but also damages an organization's reputation (Oladoyinbo et al., 2024).

6 Model Interpretation and Transparency: Key for Replication

The importance of model interpretation and transparency is rapidly increasing to meet the demands of replicability in the field of AI. Transparency refers to "making the AI system's decision-making process and underlying algorithms available to and intelligible by all parties involved, including users, developers, and regulators" (Shukla & Shukla, 2023). That is to say, transparency can help users make sense of AI systems and understand why they generate certain outcomes. On the other hand, interpretation refers to "the act of taking an explanation and transforming it into a new and clearer explanation, aimed at enhancing understanding" (Erasmus et al., 2021). Interpretation and transparency serve as main principles that help AI scientists replicate their work, which is essential for research projects. A comprehensive assessment of foundation models' transparency of major AI tools based on a total of hundred indicators ranging from data, algorithms, labor, and risks to usage policies and feedback reveals that the overall transparency has been increasing, and more developers are readily opening the "black box" of their AI tools (Bommasani et al., 2024).

Transparency plays a crucial role in allowing researchers to explore the interaction between data and algorithms. Understanding AI models and how they function is necessary to make judgments about the predictions they make and the biases in the data itself. These issues are becoming essential, especially in the medical domain, where mistakes in decisions can lead to serious consequences. In one study, Rudin (2019) argues that while explaining the black-box models sounds like a solution, it often does not function well in addressing some issues and can lead to serious problems. Alternatively, the author advocates the development of models to be employed in critical areas such as healthcare, computer vision, and criminal justice. In another study by Caruana et al. (2015), the authors showed that generalized additive models with pairwise interactions offer both performance and high accuracy in healthcare applications. Indeed, these approaches not only offer state-of-the-art accuracy, but they can also be explored for numerous critical applications in many domains, such as those in health and computer vision, where high accuracy is key.

In addition to these benefits, transparency is vital for ensuring that AI models can be reliably reproduced, which is essential for validating and building upon research findings. Without providing clarity of model interpretation, it can be harder for researchers to reproduce models. Kocak et al. (2023) emphasize that achieving reproducibility in AI research requires close attention from authors, reviewers, and journal editors to ensure transparency, along with the availability of data, code, and models in future studies. Ensuring this level of openness not only builds trust in the process but also promotes collaboration among researchers from various disciplines. By offering more accessibility to study elements, researchers can make rapid advancements in the field. Additionally, recent research in AI models designed to predict the replicability of research papers provides promising outcomes to enhance transparency in the field (Yang et al., 2020). These models that have delivered high accuracy in identifying outcomes likely to replicate highlight the importance of combining human judgment with AI to enhance the reliability of scientific findings across disciplines.

In the context of data interpretation, Generative AI (GenAI) models, such as ChatGPT, can play a significant role. These models can analyze large amounts of text-based data and provide summaries, explanations, and insights. For example, when researchers use ChatGPT to analyze complex data, it offers a wide range of techniques for coding and qualitative data interpretation (Sen et al., 2023). ChatGPT has the ability to streamline data analysis and make it useful for researchers who need to work efficiently while assuring that the core messages and patterns of the data are accurately understood and represented (Sen et al., 2023). However, it is essential to combine human judgment and experience with these AI interpretations to ensure that biases are addressed (Kitamura, 2023).

Although sophisticated AI models such as deep neural networks can be thought of as too complex to be fully interpreted, it is essential to maintain a balance between complexity and transparency. There is a growing demand to build hybrid approaches that have a blend of interpretable models and high-performance algorithms. This is evidenced by the work of Choi et al. (2016), who developed the REverse Time AttentIoN model (RETAIN). RETAIN has an effective combination of high accuracy and clinical interpretability as it employs a two-level attention mechanism to focus on essential clinical data in a way that mirrors physician practice. Their approach shows how hybrid models can bridge the gap between accuracy and interpretability and, therefore, offer a promising path for applications where it is essential to maintain a blend between the two elements. These models can be highly effective in real-world clinical environments.

Moreover, when applying transparency principles to GenAI tools (e.g., ChatGPT), researchers should emphasize the importance of model auditing and regular evaluations (Birtill & Birtill, 2024; Huang al., 2024; Thiers & Lucy, 2024). For instance, ChatGPT might generate different interpretations depending on subtle variations in the input, reflecting how its underlying algorithms process data in real time (Amaro et al., 2023). This makes it crucial to consider transparency in these AI-driven models, ensuring investigators can identify the sources of specific outcomes and make the needed adjustments. The insights related to how GenAI interprets data and outcomes help AI practitioners further refine these tools to be applied in critical areas such as law and healthcare.

From another perspective, the relationship between transparency and bias in AI is essential to ensure fairness and accountability. Transparency can lead to identifying biases embedded in AI systems by adding more visibility to the algorithms, which allows for identifying and correcting unfair practices. Strategies such as Explainable AI can be implemented to enhance the clarity of the models to help identify biases (Akinrinola et al., 2024). This visibility is essential to address ethical concerns and maintain fairness in AI applications by facilitating monitoring and continuous improvements. Furthermore, as noted by Abbu et al. (2022), the credibility of AI results relies not only on technological innovations but also on maintaining trust via ethical practices. Hence, it is essential for successful digital leaders to integrate human-centered strategies to ensure the fairness and reliability of AI tools, aligning with ethical frameworks to promote confidence in results.

It is worth noting that model interpretation and any conclusions that follow must consider the technical performance of any given model, clearly understanding a model's shortcomings in each specific context and regarding specific tasks it is meant to complete. While some tasks have seen AI tools equal or surpass human performance, in others, such as moral reasoning, they are not fully aligned with human judgment (Maslej et al., 2024), and, hence, can't be relied upon for decision-making.

In summary, it is essential to incorporate model interpretation and transparency to advance AI research. Ensuring that AI systems are transparent and reproducible is fundamental to building trust across different fields. Additionally, addressing bias via transparent practices enhances fairness and promotes ethical responsibility in the development of AI systems.

7 Case Studies and Best Practices

Considering the ubiquitousness of AI technologies in today's society and it potential to have both negative and positive impact on human wellbeing, the following five principles for ethical AI have been identified based on the current literature and prominent societal initiatives by Floridi and Cowls (2022). These principles are beneficence, nonmaleficence, autonomy, justice and explicability. The principle of beneficence refers to promoting well-being, preserving dignity and sustaining the planet building directly on Aristotelian and Kantian ethics while also incorporating the principles of sustainable development (Elkington, 2004). The principle of nonmaleficence refers to the basic notion of 'do no harm' and emphasizes the importance of personal privacy and security cautioning against risks from technological innovation. The principle of autonomy echoing the fundamental right to human autonomy proclaimed by Johnson and Cureton (2004) focuses on the power to decide as the main prerogative that should remain in the hands of humans who should decide what to delegate and not to the machines. The principle of justice is meant to promote prosperity, preserving solidarity and avoiding unfairness striving to avoid past injustice and eliminate any form of discrimination. Lastly, the principle of explicability should enable other principles through intelligibility and accountability of the decision-makers and parties involved in the process.

From a practical perspective, data for AI research should follow similar best practices as for other types of research but emphasize data context in more detail to ensure that the settings and data are appropriate for a given research question, be fully transparent regarding the methodology used to obtain or generate training data, ensuring anonymity while also allowing replication that is crucial for quality scientific research (Grégoire et al., 2024). The question of transparency is of outmost importance for assessing reliability and validity of conclusions obtained using AI tools (Grimes et al., 2023).

When it comes to synthetic or simulated datasets, multiple techniques have been used in literature, e.g., in cybersecurity field using protocol fuzzing-based methodology to generate the network traffic data (Zou et al., 2022), generative adversarial networks for creating synthetic attack data (Zhao et al., 2024), or synthetic minority over-sampling technique for additional sampling of rarer language cases used for training in sentiment analysis (El-Alfy & Al-Azani, 2020). Silicon samples can also be used for measure development and item pre-testing, and pilot testing in consumer research as they are less costly and allow for time savings (Goh et al., 2023; Sarstedt et al., 2024). Sarstedt et al. (2024) provide detailed recommendations on how to best use generative AI for such purposes. First of all, the researchers must evaluate whether the available training data have relevant information to adequately train the algorithm. Next, it is important to properly customize training data and take time for prompt-tuning, i.e., an iterative process of customizing prompts. Additionally, it is necessary to benchmark constructs generated with a help of a silicon sample against real data available from prior research to establish its reliability as well as convergent, discriminant and nomological validity (Goh et al., 2023). Ultimately, for the sake of transparency and research integrity, it is important to keep record of all the model settings and parameters so other research may reproduce a study design in the future. Needless to say, that use of a silicon sample must always be disclosed and never presented as a real data collected from a real sample despite the easy of create fake datasets with the help of generative AI as ChatGPT (Fiore, 2023).

Tables 9.2 and 9.3 provide several examples drawn from multiple disciplines of how AI research is done and categorizes these examples based on the best practices identified in the literature and presented in this chapter. The table provides an overview of the contemporary practices in data collection along with the ethical practices across various AI studies. It is essential for AI researchers to understand the differences between the methods to ensure the replicability and reliability of their

Table 9.2 Overview of data collection practices in AI research (part 1)

Authors & year	Journal	Topic	Design	Use of AI researchers	Data collection context explained and justified
Van den Broek et al. (2021)	MIS Quarterly	Machine learning vs. export assessment in hiring process	Ethnography	Indirect (by the company studied)	Yes
Al-Jamaan et al. (2023)	Journal of Intelligent and Fuzzy Systems	Deep learning methods for geo-location prediction	Empirical	Direct	Yes
Faheem and Dutta (2023)	Journal of Knowledge Management	IBM Watson failure for medical diagnostic	Case study	Indirect (by the company studied)	Yes
Ashwin Shenoy and Thillaiarasu (2023)	Journal of Intelligent and Fuzzy Systems	Machine learning for human activity recognition	Empirical	Direct	Context explained, temple setting questionable
Tian et al. (2024)	Journal of Marketing Research	Machine learning for influencing selection in social media (TikTok)	Empirical	Direct	Yes

research. Additionally, the table shows the diverse methods that are employed in data handling, including the informed consent and the preservation of anonymity. While some studies offer a clear view of consent practices, some others have less details. These variations can be reminders of the ethical responsibilities that are related to AI. In general, the overview can help AI researchers refine their own practices to ensure that they adhere to the highest ethical standards and contribute to the AI community.

8 Conclusion

While AI offers rich data collection and processing capabilities that have gained widespread use in both business and academic research, challenges linked to data integrity should not be overlooked throughout each step of the research design to ensure rigor and replicability. From broad data governance to the intricacies of data collection methods, data interpretation and transparency, multiple constraints must be taken into account to ensure data integrity. Moreover, ethical considerations with regards to potential risks to individual users' or subjects' privacy and security in

Table 9.3 Ethical considerations in AI research

Authors & year	Transparency about how & when data was obtained	Details about what data was used for training	Training dataset availability	Informed consent	Anonymity preserved
Van den Broek et al. (2021)	Yes	Yes, included employee personality traits and performance	No	Mast likely yes due to direct participation in AI games	Unclear
Al-Jamaan et al. (2023)	Yes	Yes, twitter users geolocation, algorithm included	Shared publicly	Unspecified	Unclear
Faheem and Dutta (2023)	No	Yes, included patient medical records	No	Unspecified	Unclear
Ashwin Shenoy and Thillaiarasu (2023)	Yes	Yes, algorithm included	No	Yes	Unclear
Tian et al. (2024)	Yes	Yes	No	Unspecified	Unclear

addition to legal implications of research design choices should also be given proper attention. Ultimately, striving to follow the five principles articulated by Floridi and Cowls (2022), i.e., principles are beneficence, nonmaleficence, autonomy, justice and explicability, will help avoiding common ethical pitfalls associated with data integrity in AI research. Although general guidelines related to data collection and handling in AI studies follow standard recommendations of scientific studies as upheld by IRB, the sheer amount of data available and handled in AI context adds to the design complexity and creates additional constraints to consider to avoid causing unintentional damage both at the data collection and handling stages and the outcomes of the decisions taken based on the algorithms trained on such data.

References

Abbu, H., Mugge, P., & Gudergan, G. (2022). Ethical considerations of artificial intelligence: Ensuring fairness, transparency, and explainability. In 2022 Ieee 28th international conference on engineering, technology and innovation (ice/itmc) & 31st international association for management of technology (iamot) joint conference (pp. 1–7).

Abraham, R., Schneider, J., & Vom Brocke, J. (2019). Data governance: A conceptual framework, structured review, and research agenda. *International Journal of Information Management, 49*, 424–438.

Ahmadi, S., Tavana, M. M., Shokouhyar, S., & Dortaj, M. (2022). A new fuzzy approach for managing data governance implementation relevant activities. *The TQM Journal, 34*(5), 979–1012.
Akinrinola, O., Okoye, C. C., Ofodile, O. C., & Ugochukwu, C. E. (2024). Navigating and reviewing ethical dilemmas in AI development: Strategies for transparency, fairness, and accountability. *GSC Advanced Research and Reviews, 18*(3), 050–058.
Al-Jamaan, R., Ykhlef, M., & Alothaim, A. (2023). Gldm: Geo-location prediction of twitter users with deep learning methods 1. *Journal of Intelligent & Fuzzy Systems, 45*(2), 2723–2734.
Amaro, I., Della Greca, A., Francese, R., Tortora, G., & Tucci, C. (2023). Ai unreliable answers: A case study on chatgpt. In International conference on human-computer interaction (pp. 23–40).
Araújo, S. C., Salles, P., & Saito, C. H. (2008). A case study on qualitative model evaluation using data about river water quality. *Ecological Informatics, 3*(1), 13–25.
Ashwin Shenoy, M., & Thillaiarasu, N. (2023). Enhancing temple surveillance through human activity recognition: A novel dataset and yolov4-convlstm approach. *Journal of Intelligent & Fuzzy Systems (Preprint)*, 1–16.
Bhattacherjee, A. (2012). *Social science research: Principles, methods, and practices*. University of South Florida.
Birtill, M., Birtill, P. (2024) Implementation and evaluation of genai-aided tools in a uk further education college. In Artificial Intelligence Applications in Higher Education, pp. 195–214. Routledge.
Bommasani, R., Klyman, K., Longpre, S., Kapoor, S., Maslej, N., Xiong, B., ... Liang, P. (2024). The foundation model transparency index. Online. Retrieved from https://crfm.stanford.edu/fmti/ May-2024/index.html (Retrieved on September 5, 2024).
Buhmann, A., & Fieseler, C. (2023). Deep learning meets deep democracy: Deliberative governance and responsible innovation in artificial intelligence. *Business Ethics Quarterly, 33*(1), 146–179.
Caci, B., & Dhou, K. (2020). The interplay between artificial intelligence and users' personalities: A new scenario for human-computer interaction in gaming. In HCI international 2020–late breaking papers: Cognition, learning and games: 22nd HCI international conference, hcii 2020, Copenhagen, Denmark, July 19–24, 2020, proceedings (pp. 619–630).
Caruana, R., Lou, Y., Gehrke, J., Koch, P., Sturm, M., & Elhadad, N. (2015). Intelligible models for healthcare: Predicting pneumonia risk and hospital 30-day readmission. In Proceedings of the 21th ACM SIGKDD international conference on knowledge discovery and data mining (pp. 1721–1730).
Choi, E., Bahadori, M. T., Sun, J., Kulas, J., Schuetz, A., & Stewart, W. (2016). Retain: An interpretable predictive model for healthcare using reverse time attention mechanism. *Advances in Neural Information Processing Systems, 29*.
Daniel, D. A., & Roslin, S. E. (2021). Data validation and integrity verification for trust-based data aggregation protocol in WSN. *Microprocessors and Microsystems, 80*, 103354. Retrieved from https://www.sciencedirect.com/science/article/pii/S0141933120305135. https://doi.org/10.1016/j.micpro.2020.103354
Dhou, K. (2018a). A novel agent-based modeling approach for image coding and lossless compression based on the wolf-sheep predation model. In Computational science–ICCS 2018: 18th international conference, Wuxi, China, June 11-13, 2018, proceedings, part II 18 (pp. 117–128).
Dhou, K. (2018b). Towards a better understanding of chess players' personalities: A study using virtual chess players. In Human-computer interaction. Interaction technologies: 20th international conference, HCI international 2018, Las Vegas, NV, USA, July 15–20, 2018, proceedings, part III 20 (pp. 435–446).
Dhou, K. (2019). An innovative employment of virtual humans to explore the chess personalities of Garry Kasparov and other class-A players. In HCI international 2019–late breaking papers: 21st HCI international conference, HCII 2019, Orlando, FL, USA, July 26–31, 2019, proceedings 21 (pp. 306–319).
Dhou, K. (2020a). A new chain coding mechanism for compression stimulated by a virtual environment of a predator-prey ecosystem. *Future Generation Computer Systems,*

102, 650–669. Retrieved from https://www.sciencedirect.com/science/article/pii/S0167739X1832630X. https://doi.org/10.1016/j.future.2019.08.021

Dhou, K. (2020b). A novel investigation of attack strategies via the involvement of virtual humans: A user study of Josh Waitzkin, a virtual chess grandmaster. In HCI international 2020–late breaking papers: Cognition, learning and games: 22nd HCI international conference, HCII 2020, Copenhagen, Denmark, July 19–24, 2020, proceedings 22 (pp. 658–668).

Dhou, K. (2021a). An exploration of chess personalities in grandmasters and class-A players using virtual humans. *International Journal of Entertainment Technology and Management, 1*(2), 126–145. Retrieved from https://www.inderscienceonline.com. https://doi.org/10.1504/IJENTTM.2021.116767

Dhou, K. (2021b). An exploration of the fear of attack strategy in chess and its influence on class-A players of different chess personalities: An exploration using virtual humans. In international conference on human-computer interaction (pp. 185–195).

Dhou, K. (2023). An exploration of feared versus fearless attack attitudes using the chess personalities of virtual chess players. In International conference on human-computer interaction (pp. 176–185).

Dhou, K., & Cruzen, C. (2020). A new chain code for bi-level image compression using an agent-based model of echolocation in dolphins. In 2020 IEEE 6th international conference on dependability in sensor, cloud and big data systems and applications (DependSys) (pp. 87–91).

Dhou, K., & Cruzen, C. (2021a). A highly efficient chain code for compression using an agent-based modeling simulation of territories in biological beavers. *Future Generation Computer Systems, 118*, 1–13. Retrieved from https://www.sciencedirect.com/science/article/pii/S0167739X20330788. https://doi.org/10.1016/j.future.2020.12.016

Dhou, K., & Cruzen, C. (2021b). An innovative employment of the NetLogo AIDS model in developing a new chain code for compression. In International conference on computational science (pp. 17–25).

Dhou, K., & Cruzen, C. (2022a). A creative chain coding technique for bi-level image compression inspired by the NetLogo HIV agent-based modeling simulation. *Journal of Computational Science, 61*, 101613. Retrieved from https://www.sciencedirect.com/science/article/pii/S1877750322000394. https://doi.org/10.1016/j.jocs.2022.101613

Dhou, K., & Cruzen, C. (2022b). A hypothetical agent-based model inspired by the abstraction of solitary behavior in tigers and its employment as a chain code for compression. In International conference on computational science (pp. 96–102).

Dhou, K., & Cruzen, C. (2022c). An innovative chain coding mechanism for information processing and compression using a virtual bat-bug agent-based modeling simulation. *Engineering Applications of Artificial Intelligence, 113*, 104888. Retrieved from https://www.sciencedirect.com/science/article/pii/S0952197622001282. https://doi.org/10.1016/j.engappai.2022.104888

DoCarmo, T., Rea, S., Conaway, E., Emery, J., & Raval, N. (2021). The law in computation: What machine learning, artificial intelligence, and big data mean for law and society scholarship. *Law & Policy, 43*(2), 170–199.

Draghi, B., Wang, Z., Myles, P., & Tucker, A. (2024). Identifying and handling data bias within primary healthcare data using synthetic data generators. *Heliyon, 10*(2).

Durmus, E., Nyugen, K., Liao, T. I., Schiefer, N., Askell, A., & Bakhtin, A., et al. (2023). Towards measuring the representation of subjective global opinions in language models. arXiv e-prints, arXiv:2306.

El-Alfy, E.-S. M., & Al-Azani, S. (2020). Empirical study on imbalanced learning of Arabic sentiment polarity with neural word embedding. *Journal of Intelligent & Fuzzy Systems, 38*(5), 6211–6222.

Elkington, J. (2004). In A. Henriques & J. Richardson (Eds.), *The triple bottom line: Does it all add up?* (pp. 23–38). Earthscan.

Erasmus, A., Brunet, T. D., & Fisher, E. (2021). What is interpretability? *Philosophy & Technology, 34*(4), 833–862.

Faheem, H., & Dutta, S. (2023). Artificial intelligence failure at IBM's Watson for oncology. *IUP Journal of Knowledge Management, 21*(3), 47–75.

Fan, C., Esparza, M., Dargin, J., Wu, F., Oztekin, B., & Mostafavi, A. (2020). Spatial biases in crowdsourced data: Social media content attention concentrates on populous areas in disasters. *Computers, Environment and Urban Systems, 83*, 101514.

Ferrara, E. (2023). Fairness and bias in artificial intelligence: A brief survey of sources, impacts, and mitigation strategies. *Science, 6*(1), 3.

Fiore, K. (2023, November). ChatGPT created a fake dataset with skewed results. MedPage Today. Retrieved from https://www.medpagetoday.com/special-reports/features/107247.

Floridi, L., & Cowls, J. (2022). In S. Carta (Ed.), *Machine learning and the city: Applications in architecture and urban design* (pp. 535–545). Wiley Online Library.

Fung, B. (2024, June). What is Apple's AI doing with your data? CNN.com. Retrieved from https://www.cnn.com/2024/06/13/tech/apple-ai-data-openai-artificial-intelligence/index.html.

George, R., Truong, T., & Davidson, J. (2017). Establishing an effective data governance system: Data governance is necessary for compliance with current regulatory expectations or data integrity in pharmaceutical R&D and manufacturing organizations. *Pharmaceutical Technology Europe, 29*(11), 40–44.

Gilbert, A. (2024). Nvidia illegally scraped YouTube videos for AI, suit says. Bloomberg Law. Retrieved from https://news.bloomberglaw.com/ip-law/nvidia-illegally-scrapped-youtube-videos-to-train-ai-suit-says.

Goh, T.-T., Dai, X., & Yang, Y. (2023). Benchmarking ChatGPT for prototyping theories: Experimental studies using the technology acceptance model. *BenchCouncil Transactions on Benchmarks, Standards and Evaluations, 3*(4), 100153. Retrieved from https://www.sciencedirect.com/science/article/pii/S277248592400005X. https://doi.org/10.1016/j.tbench.2024.100153

Grant, N. (2024). Google takes the next step in its AI evolution. The New York Times. Retrieved from https://www.nytimes.com/2024/05/14/technology/google-artificial-intelligence-products.html.

Grégoire, D. A., Ter Wal, A. L., Little, L. M., Bermiss, S., Kotha, R., & Gruber, M. (2024). Mobilizing new sources of data: Opportunities and recommendations. *Academy of Management Journal, 67*(2), 289–298.

Grimes, M., Von Krogh, G., Feuerriegel, S., Rink, F., & Gruber, M. (2023). From scarcity to abundance: Scholars and scholarship in an age of generative artificial intelligence. *Academy of Management Journal, 66*(6), 1617–1624.

Helmrich, A. M., Ruddell, B. L., Bessem, K., Chester, M. V., Chohan, N., Doerry, E., et al. (2021). Opportunities for crowdsourcing in urban flood monitoring. *Environmental Modelling & Software, 143*, 105124.

Huang, K., Yeoh, J., Wright, S., & Wang, H. (2024). Build your security program for GenAI. In *Generative AI security: Theories and practices* (pp. 99–132). Springer.

Islam, G., & Greenwood, M. (2024). Generative artificial intelligence as Hypercommons: Ethics of authorship and ownership. *Journal of Business Ethics*, 1–5.

Janssen, M., Brous, P., Estevez, E., Barbosa, L. S., & Janowski, T. (2020). Data governance: Organizing data for trustworthy artificial intelligence. Government information quarterly, 37(3), 101493.

Jauhiainen, T., & Lehner, O. M. (2022). Good governance of AI and big data processes in accounting and auditing. In Artificial intelligence in accounting (pp. 119–181). Routledge.

Jiang, Y., Jeusfeld, M., & Ding, J. (2021). Evaluating the data inconsistency of open-source vulnerability repositories. In Proceedings of the 16th international conference on availability, reliability and security (pp. 1–10).

Jiménez, J. (2024). Can I opt out of Meta's A.I. scraping on Instagram and Facebook? Sort of. The New York Times. Retrieved from https://www.nytimes.com/2024/06/07/technology/meta-ai-scraping-policy.html.

Jobin, A., Ienca, M., & Vayena, E. (2019). The global landscape of AI ethics guidelines. *Nature Machine Intelligence, 1*(9), 389–399.

Johnson, R., & Cureton, A. (2004). Kant's moral philosophy.

Jones, B. (2023). How generative AI tools help transform academic research. Forbes. Retrieved from https://www.forbes.com/sites/beatajones/2023/09/28/how-generative-ai-tools-help-transform-academic-research/.

Kirova, V. D., Ku, C., Laracy, J., & Marlowe, T. (2023). The ethics of artificial intelligence in the era of generative AI. *Journal of Systemics, Cybernetics and Informatics, 21*(4), 42–50.

Kitamura, F. C. (2023). ChatGPT is shaping the future of medical writing but still requires human judgment. *Radiological Society of North America, 307*(2).

Kocak, B., Yardimci, A. H., Yuzkan, S., Keles, A., Altun, O., Bulut, E., et al. (2023). Transparency in artificial intelligence research: A systematic review of availability items related to open science in radiology and nuclear medicine. *Academic Radiology, 30*(10), 2254–2266.

Lee, M. (2023). Recent advances in generative adversarial networks for gene expression data: A comprehensive review. *Mathematics, 11*(14), 3055.

Lu, N., Zhang, Y., Shi, W., Kumari, S., & Choo, K.-K. R. (2020). A secure and scalable data integrity auditing scheme based on Hyperledger fabric. *Computers & Security, 92*, 101741.

Maaliw, R. R., III, & Degadwala, S. (2023). Artificial intelligence and blockchain-based optimized and fair payment scheme over cloud computing system. *Journal of Engineering Science and Mathematics*, 22.

Maher, N. A., Senders, J. T., Hulsbergen, A. F., Lamba, N., Parker, M., Onnela, J.-P., et al. (2019). Passive data collection and use in healthcare: A systematic review of ethical issues. *International Journal of Medical Informatics, 129*, 242–247.

Marcus, G. F. (2024). *Taming Silicon Valley: How we can ensure that AI works for us*. MIT Press.

Maslej, N., Fattorini, L., Perrault, R., Parli, V., Reuel, A., Brynjolfsson, E., et al. (2024). *The AI index 2024 annual report*. Tech. Rep. Stanford University.

Morey, T., Forbath, T., & Schoop, A. (2015). Customer data: Designing for transparency and trust. *Harvard Business Review, 93*(5), 96–105.

Mouring, M., Dhou, K., & Hadzikadic, M. (2018). A novel algorithm for bi-level image coding and lossless compression based on virtual ant colonies. In The 3rd international conference on complexity, future information systems, and risk (pp. 72–78).

Nasr, M., Carlini, N., Hayase, J., Jagielski, M., Cooper, A., & Ippolito, D., et al. (2017). Scalable extraction of training data from (production) language models. arXiv preprint arXiv:2311.17035.

Oladoyinbo, T. O., Olabanji, S. O., Olaniyi, O. O., Adebiyi, O. O., Okunleye, O. J., & Ismaila Alao, A. (2024). Exploring the challenges of artificial intelligence in data integrity and its influence on social dynamics. *Asian Journal of Advanced Research and Reports, 18*(2), 1–23.

Pansara, R. (2023). Unraveling the complexities of data governance with strategies, challenges, and future directions. *Transactions on Latest Trends in IoT, 6*(6), 46–56.

Poth, A., Wildegger, A., & Levien, D.-A. (2024). Considerations about integration of GenAI into products and services from an ethical and legal perspective. In European conference on software process improvement (pp. 155–171).

Redman, T. C., & Hoerl, R. W. (2024). AI and statistics: Perfect together. *MIT Sloan Management Review*, 1–4.

Rezig, E. K., Ouzzani, M., Elmagarmid, A. K., Aref, W. G., & Stonebraker, M. (2019). Towards an end-to-end human-centric data cleaning framework. In Proceedings of the workshop on human-in-the-loop data analytics (pp. 1–7).

Rudin, C. (2019). Stop explaining black box machine learning models for high stakes decisions and use interpretable models instead. *Nature Machine Intelligence, 1*(5), 206–215.

Sarstedt, M., Adler, S. J., Rau, L., & Schmitt, B. (2024). Using large language models to generate silicon samples in consumer and marketing research: Challenges, opportunities, and guidelines. *Psychology & Marketing, 41*(6), 1254–1270. Retrieved from https://onlinelibrary.wiley.com/doi/abs/10.1002/mar.21982. https://doi.org/10.1002/mar.21982

Sen, M., Sen, S. N., & Sahin, T. G. (2023). A new era for data analysis in qualitative research: ChatGPT. *Shanlax International Journal of Education, 11*, 1–15.

Shukla, R., & Shukla, A. (2023). Ethical considerations in artificial intelligence: Ensuring accountability and transparency. *IRE Journals*.

Singhal, S., Kothuru, S. K., Sethibathini, V. S. K., & Bammidi, T. R. (2024). ERP excellence a data governance approach to safeguarding financial transactions. *International Journal of Management Education for Sustainable Development, 7*(7), 1–18.

Sosko, S., & Dalyot, S. (2017). Crowdsourcing user-generated mobile sensor weather data for densifying static geosensor networks. *ISPRS International Journal of Geo-Information, 6*(3), 61.

Thiers, F. A., & Lucy, K. (2024). *A distinct approach to clinical GenAI oversight*. Tech. Rep. Center for Open Science.

Thongmeensuk, S. (2024). Rethinking copyright exceptions in the era of generative AI: Balancing innovation and intellectual property protection. *The Journal of World Intellectual Property, 27*(2), 278–295.

Tian, Z., Dew, R., & Iyengar, R. (2024). Mega or micro? Influencer selection using follower elasticity. *Journal of Marketing Research, 61*(3), 472–495.

Ugboma, E. (2004). Assuring information systems' effectiveness through data integrity: Essential guidelines for information systems databases. In The proceedings of ISECON (Vol. 21, pp. 1–6).

Van den Broek, E., Sergeeva, A., & Huysman, M. (2021). When the machine meets the expert: An ethnography of developing AI for hiring. *MIS Quarterly, 45*(3).

Voola, P. K., Ayyagiri, A., Musunuri, A., Aggarwal, A., & Jain, S. (2024). Leveraging GenAI for clinical data analysis: Applications and challenges in real-time patient monitoring. *Modern Dynamics: Mathematical Progressions, 1*(2), 204–223.

Walmsley, J. (2021). Artificial intelligence and the value of transparency. *AI & Society, 36*(2), 585–595.

Xie, K., Ning, X., Wang, X., Xie, D., Cao, J., Xie, G., & Wen, J. (2016). Recover corrupted data in sensor networks: A matrix completion solution. *IEEE Transactions on Mobile Computing, 16*(5), 1434–1448.

Yang, Y., Youyou, W., & Uzzi, B. (2020). Estimating the deep replicability of scientific findings using human and artificial intelligence. *Proceedings of the National Academy of Sciences, 117*(20), 10762–10768.

Zhao, X., Fok, K. W., & Thing, V. L. (2024). Enhancing network intrusion detection performance using generative adversarial networks. *Computers & Security, 145*, 104005. Retrieved from https://www.sciencedirect.com/science/article/pii/S0167404824003109. https://doi.org/10.1016/j.cose.2024.104005

Zou, Q., Singhal, A., Sun, X., & Liu, P. (2022). Deep learning for detecting logic-flaw-exploiting network attacks: An end-to-end approach. *Journal of Computer Security, 30*(4), 541–570.

Part IV
Policy Perspectives and Future Directions in AI and Higher Education

Chapter 10
AI Policy Considerations in Academic Institutions

Tasnim Ahmed and Olatunde Durowoju

Abstract Higher Education (HE), globally, is facing significant transformations due to the heavy influence of Artificial Intelligence (AI) applications within the EdTech industry. This chapter discusses the complex opportunities and challenges presented by the adoption of such technologies within the HE. It argues that, inevitably, the usage of such technologies including generative AI (such as ChatGPT), provides assurances to enhance pedagogy, student support and facilitate research activities. Despite such assurances, there are challenges with its usage, including potential biases, academic integrity and sustainability. The various measures around the globe are discussed as countries try to regulate and monitor these fast-paced technologies, depending on their socio-economic, political and cultural practices. Finally, this chapter advocates for solutions to integrate AI within the HE sector through a potential roadmap, whilst upholding ethics, equity and non-discrimination.

Keywords AI policy · Higher education · Academic integrity · Ethics · Inclusion

1 Introduction

The latest development within machine learning, natural language processing and data analytics have increased the potential of Artificial Intelligence (AI) to transform the global higher education sector (Rudolph et al., 2024). The emergence of generative AI (GAI) has allowed the production of texts, images videos from prompts and the use of AI in education is growing rapidly to meet the demands of businesses who encourage the deployment of AI technologies within their respective industries. However, ethical risks persist, and universities are challenged with

T. Ahmed · O. Durowoju (✉)
Liverpool John Moores University, Liverpool, UK
e-mail: O.A.Durowoju@ljmu.ac.uk

X. Zhou, H. Al-Samarraie (eds.), *Institutional guide to using AI for research*, Advances in Artificial Intelligence in Education 2,
https://doi.org/10.1007/978-3-031-94809-1_10

tackling bias and possible unfairness stemming from the use of AI-powered tools whilst maintaining their regular operations such as teaching and learning, research and outreach and engagement (Hijazi, 2024). Undoubtedly, AI tools can provide students with greater access to education allowing inclusivity for all students. It is acknowledged that AI may offer students personalised learning services specific to their individual needs; thereby creating learning pathways to accommodate learning on a one-to-one basis (Onesi-Ozigagun et al., 2024). This is achieved by providing instant materials such as article summaries, essay writing and explaining concepts. Teaching methodologies can be revolutionised as AI applications provide a range of analytical functions to include patterns on student progression, assisting tutors to vary their instructional strategies (Chaudhry & Kazim, 2022). However, higher education (HE) institutions need to devise plans for the use of future development of AI tools as they are required to maintain a balance between providing their students with access to innovative technologies and yet ensuring that students play fairly and are not merely producing AI generated work under their own name. Universities have a duty to foster the use of AI in a socially responsible manner. This includes promoting AI tools which endorse equity, social justice and bias mitigation, alongside the continual assessment of their strategies to support the use of such transformative technologies. This chapter will explore the strengths of AI tools usage and offset this by considering the challenges set by AI tools in education ranging from AI usage in research, potential biases, to security and data protection. The discussion will proceed to present the rationale of an AI education policy to monitor the ethical usage of such tools in education. Current AI institutional policies will be discussed utilising national and international contexts. Finally, recommendations will be made to develop a roadmap on AI education policy as from the education perspective the use of AI tools within teaching and learning carry with themselves gaps and uncertainties. There is a need for appropriate AI policies to embed generative AI within education, whilst ensuring that privacy, security, environmental, ethical and regulatory guidelines are upheld. This needs to be conducted with continuous human monitoring and critical evaluation.

2 An Insight Within the Literature on AI Education Policy

Academic scholarship points out that continuous amendments to regulatory frameworks are necessary for responding to upcoming academic cultures envisaged through the challenges presented by generative AI (Glendinning, 2022). However, there is limited debate on the emergence of AI policies within higher education settings. Chan (2023) gathered data from staff and students across institutions reflecting upon their opinions on embedding AI policy within institutions. Whilst Chan proposed model realigns with the UNESCO AI education policy, promoting input from all stakeholders including staff, students and the leadership (UNESCO, 2021). It does not delve into the academic integrity challenges faced by international students (Bannister et al., 2024). Xiao et al. (2023) collated data from 500 universities with top QS rankings and found that a mere quarter of these institutions had generative AI policies related to generative AI impacts on academic integrity, although two thirds of the institutions advocated for monitored usage of generative AI rather than an outright ban. Perkins and Roe (2023) analysed the publicly available academic integrity policies on generative AI for 142 higher education institutions (HEIs) and reported significant gaps on generative AI's impacts on academic integrity. Hence there was the need to devise transparent policies which integrate education and AI whilst allowing higher institutions to take advantage of the facilities offered by AI applications (Perkins & Roe, 2023). A further problem highlighted was that generative AI contains certain biases against non-native English users. This requires urgent changes to policy frameworks that contend with the academic integrity dilemmas faced by global students (Liang et al., 2023). In the absence of limited scholarly attention provided to the governance of generative AI academic integrity this exploratory study aims to examine the higher institutions generative AI academic integrity policy landscape to offer policy initiatives for differing student cohorts present on campuses.

The research questions are as follows: i) To what extent have academic integrity policies created by HEIs made provisions to ensure inclusivity and bias mitigation within current AI application use and deployment within the HE sector. ii) Secondly, what are the challenges and the strengths presented by current policies and how can the future policy development discourse support the embedding of AI applications within the curricula whilst protecting any ethical concerns arising from such usage.

The research design, which is preliminary desktop study entailing a systematic review, consisted of taking a wider consensus from previous studies within the field and utilising publicly available web-based sources aligning with Patton et al's policy analysis methodology (Patton et al., 2017). An extensive web search and institutional database search was carried out to retrieve journal articles, books, blogs and policies from a variety of disciplines using key terms such as "AI education policy', 'generative AI and higher education', 'AI AND research AND 'higher education', 'generative AI' AND 'higher education policy'. The search was limited from 2001–2024. Following these various searches 128 publications were obtained from the institutional library catalogue search. Following the examination of the abstracts, the most relevant were utilised. The remaining online data was retrieved using these search terms on the web. This research data collection technique enabled the authors to conduct an analysis of the documents to present focal points and priorities in terms of the current AI educational policies and the challenges and opportunities posed by usage of such tools in the higher education sector.

2.1 Strengths and Challenges of Deploying AI Applications within Education

AI has undoubtedly gained a lot public attention since the introduction of large language models (LLMs) such as ChatGPT in 2023. The higher education sector's adoption of the same leads to public attention and deliberation (Bozkurt et al., 2023). Whilst gaining public attention on research can be positive it can also entail negative hype. This hype ranges from claims that AI will replace human workforce and bring in automation to replacing teachers with AI (Peters et al., 2024). To balance the 'hype' of AI in education this section of the discussion will explore the strengths and weaknesses of the embedded AI pathways within higher education. This will be followed by the rationale of introducing an AI policy to meet the demands of the Industry 4.0 era, which requires institutions to adopt dynamic policies allowing students to gain experiential learning opportunities preparing them for the modern job market.

2.1.1 Strengths

Personalised Learning

AI tools provide students with individualised learning support by evaluating their performance through quizzes and task completions. From these results the system can recommend further learning, additional resources and further tasks depending on the student's mastery level of the concepts (Usman et al., 2024). Each student is unique, and AI can adapt to accommodate student learning at their own pace. Previous studies have indicated that student performance increase motivation and engagement, which can greatly improve via customised pathways offered through personalised learning (Atadoga et al., 2024).

Metamorphosing Teaching Methodologies

AI tools are metamorphosing teaching methodologies through instructional design and student engagement. This is done by utilising AI interactive platforms, which promote student engagement. Examples of this type of engagement include multimedia stimulations and gaming styled activities which intend to make learning more immersive and encourage critical thinking (Nwokwdiegwo et al., 2024). AI teaching methodologies also encourage collaborative learning forums as AI platforms can be utilised for students to collaborate during group work, these skills are encouraged in today's workforce. AI chatbots can act as teaching assistants carrying out cognitive functions such as brainstorming and drafting of lesson plans (Mollick & Euchner, 2023).

Remodelling Assessment

AI powered assessment tools can analyse real time responses such as multiple-choice questions, allowing instructors to identify patterns in student performance and provide interventions if required. AI systems such as automated essay scoring (AES) can be utilised in providing formative feedback to essay type questions (Gardner et al., 2021), although its use in summative assessment is still dubious owing to concerns about bias replication contained within the training set used in its programming. These tools can reduce the burden of marking (Usman et al., 2024), which can be tedious with large class sizes, and free up time for educators to be more creative and purposeful towards providing robust education for students. The data obtained from AI assessment platforms can shape assessment patterns in education to include continuous assessments, which assist with monitoring student progress. However, there are ethical concerns with AI based assessments, which include questioning how fair and transparent these tools are when used for assessments. Educators need to devise mechanisms to constantly test the fairness and transparency of such platforms.

Administrative Efficiency

AI powered tools in the education sector can increase efficiency through assisting with decision making, resource allocation and within automation of routine processes. AI tools can be used to track usage of financial resources and equipment identifying areas of wastage (Ibeh et al., 2024). AI analytic dashboards can provide administrators with information on key performance indicators such as financial expenditure; identify at risk students in return improving the overall performance of the institution. AI chatbots can personalise student engagement by providing instant support to students, especially during out-of-office hours. AI tools can automate administrative procedures such as pay roll processing and procurement. This could minimise the chances of human error and allow human resources to focus on their core competencies for better strategic advantage. From a research perspective, generative AI can assist researchers with ideation, writing, data visualisation (graphs/charts etc), as permitted within the strictures of academic integrity, although due diligence needs to be applied as AI will not act as a co-author (Kooli, 2023).

2.1.2 Challenges

Data Privacy and Security

The widespread use of AI applications also carries challenges which include risks of data privacy and security. AI systems are reliant upon voluminous amounts of student data, which may be sensitive and needs protection against unauthorised access and potential misuse. Institutions require robust regulatory frameworks and data protection measures to protect the data and ensure compliance with any relevant privacy regulations (Tan, 2023). The Russell Group principles (2023) acknowledge the ethical use of generative AI and promote its equal access. Yet also highlight the concerns for the dangers of intellectual property and privacy issues stemming from the data that staff and students may enter within the application. Universities have legal obligations under various data protection legislation to protect the personal data of staff, students, visitors, and research activities. Hence to ensure compliance many universities have placed restrictions on 'restricted' or 'highly restricted' data being submitted to AI tools such as ChatGPT. For example, the University of Reading within its Data Protection and AI policy lays out the risks posed by the usage of AI on the accuracy of data and its potential adverse effects on individuals. It has set up a University Classification Policy (UCP) to allow users to assess which information falls within such categories (University of Reading, 2024). The Data Protection AI Policy raises questions that individuals need to consider before using AI tools. The questions include considering inherent bias; unexpected outcomes; impacts on individuals and monitoring of such occurrences. If individuals are considering using AI for personal and sensitive information they may need to undertake a Data Protection Impact Assessment. If risks of non-compliance are identified, then a senior risk owner needs made accountable of the same.

The UCP defines restricted information as such information that is not be made public. This would include information which may be personal, commercially sensitive or any information which has been provided on a confidential basis. Whilst highly restricted information would entail details which if disclosed may cause harm to the university or individuals (UCP, 2024). The policy provides a detailed list what could be taken as high-risk data and sensitive information, which would be deemed to be highly restricted data. The UCP lays out roles and responsibilities of staff to ensure that there are sufficient safeguards to handle such data. Line managers need to ensure that staff are appropriately trained on the data protection and the information security policies. Whilst all staff are required to report any breaches or suspected breaches or threats to information security. Despite on paper these processes seem to be aiming to manage the usage of AI whilst handling private data. Inevitably, effective audit trails need to be in place to ensure that these processes are transparent, accurate, anti-discriminatory and ensure accountability. Other Universities such as Canterbury Christ request that staff refer to the guidance for Civil Service employees when dealing with personal data (Canterbury Christ, 2023; Government Digital Service, 2025).

Algorithmic Bias

AI systems receive their training from historical systems which may contain inaccuracies and biases. If these biases are not adequately resolved, then they can manifest themselves into inequalities within the higher education sector (Baker & Hawn, 2022). Designers of such applications need to ensure that such systems are tested vigorously to ensure that unintended biases are mitigated. Despite AI tools providing the potential to enhance the learning experience, tutors have a key role in supporting students in complex environments. Teachers need training on how to interpret results acquired from such tools to ensure the benefits of AI applications are retained without any discriminatory practices against potentially vulnerable students. The AI systems need to be transparent in their decision-making, especially when these decisions have an impact on students. Whilst the Russell Group principles (2023) mention that generative AI tools produce answers which are based on societal biases. However, the guidance does not provide any solutions to combat the existence of bias when utilising such tools.

Accessibility and Equity

The AI tools need to be accessible to all students, as the risk is inherent that AI may accentuate the disparities existing between marginalised and economically stable students. Factors such as socio-economic background, digital literacy and access to the internet need to be considered to ensure that there is a balance between the reliance of technological assistance and human interaction provided to students. Generative AI may be able to support the stress and anxiety levels of staff and students and propose appropriate changes (Hashem et al., 2024). Embedding AI tools requires institutions to invest in technology infrastructure, institutions need to ensure that the potential benefits are calculated to ensure that resources are adequately allocated to maximise impacts. There are also copyright issues as prompts produce texts which if the source is not acknowledged may lead to plagiarism (Kasneci et al., 2023).

Ambiguities in Definitions of Academic Misconduct and Authenticity

The use of AI within education has initiated debates on the definition of academic authenticity and academic misconduct. There seems to be a disparity between students and tutors in their perception and usage of ChatGPT. Students refer to it as an assistive tool bearing in mind that the generated content require verification as authenticity cannot be guaranteed. Tutors on the other hand are sceptical of this AI tool as it can be used to circumvent traditional plagiarism detection tools when student present work that was not generated through own reading and information synthesis, claiming that degree outcome may not reflect actual student engagement within the degree. Some tutors also posit that AI tools discourage critical thinking in students, a core skill expected from a university degree (REF). This inconsistency in perception between the students and educators requires a nuanced approach from institutions in their policy formation and clarity-seeking procedures. It is necessary to formulate generative AI policies which promote academic integrity providing clear definitions of collusion or plagiarism versus assistance but also encourage engaging with high specification AI tools (Duah & McGivern, 2024). Recent studies have shown that comprehensive policies on academic misconduct support lower rates of academic conduct (Noorbehbahani et al., 2022). Academic misconduct is influenced by factors such as pedagogical influences which look upon the tutor's commitment to uphold academic integrity, which in turn impacts the student's outlook of academic misconduct. Other factors include the institution's stance on academic integrity, alongside environmental influences and peer perceptions of academic integrity. An international example from the George Washington University guidelines (2023) stipulates that whilst a general permission allows students to use generative AI through selective or general permission. The default position which would be in place if the instructor has not provided any guidance is that any work submitted by a student would be considered as the student's own intellectual product. Therefore, in those conditions any work that has been produced through generative AI would be considered as cheating.

The Russell Group principles (2023) confirm that institution policies clearly state to staff and students where generative AI use is permissible. However, there is a need for universities to close the divide present between policy communication and policy development. This gap has adverse implications on students as they opt to use AI tools. Without clarity of communication of policies to students, they bear the risk of making moral choices which may raise ethical concerns.

Without clear institutional policy or guidelines for students on the challenges associated with the usage of content generated by AI systems, students bear the risk of accepting hallucinated or misinformation as accurate or may misconceive biased narratives as acceptable narratives. Both cases are ethical conundrums that HEIs must dispel. This is particularly dangerous as newer versions of generative AI claim to have improved the performance of their product, though there is always a statement of caution from providers to users that their products can make mistakes. This reported improvement in performance may provide a false sense of reliance of students and instructors on the tool without the need to authenticate or validate the information. Higher institutions need to rapidly address such liabilities via robust policies to ensure that generative AI is adopted in academia after such ethical implications have been accounted for (OpenAI, 2023).

2.2 AI and Research

Machine learning algorithms can free up time for researchers through providing analysis of content; any visible gaps present in current research and administer analysis of the data. Allowing researchers to concentrate their efforts on the theoretical and conceptual aspects of the project (Burger et al., 2023). The advantage of the natural language algorithms is that they dismantle narratives to particular data points, thereby reframing the existing knowledge. Similarly, they function as heuristic devices allowing the development of new trends and paradigm shifts (Chinnasmy et al., 2022). Through automation natural language modelling AI tools can generate text, which may act as a starting point both for novice and experienced writers, alongside saving time and costs (Alqahtani et al., 2023). The academic writing process may benefit from AI tools assisting the researcher to provide academic rigour through laying out constructive arguments. Butson and Spronken-Smith (2024) contend that generative AI transforms how we perceive knowledge, data and research problems.

However, within research AI tools bring a ‘pandora’s box’ of issues, which are pivotal to the discourse (Christou, 2024). These include, as mentioned earlier, data privacy and potential bias, but also the issue of consent. Machine learning algorithms allow data to be used repeatedly which raise questions about whether the participants had consented for the multiple use of data. In addition, there are concerns that machine learning prefers quantitative over qualitative data, this is concerning as the black box nature of the AI tools means that they are not so readily transparent or accountable. These ethical challenges are key when trying to scrutinise results and trends, AI tools fail to comprehend a nuanced comprehension of human behaviour and social constructs. These shortcomings of AI make it less likely to represent causal links, which in themselves are important cornerstone within scholarly research (Butson & Spronken-Smith, 2024).

The other shortcoming of the use of AI in research would include the AI tool failing to understand the complexities of the scientific subject. Consequently, limiting the innovative potentials of the output in comparison to human generated outputs. Hence the complete reliance of an AI generative application could lead to limiting the author’s thought processes, writing style and creativity (Tozzi & Cinelli, 2021). There is also the danger that complete reliance on the AI tool may raise questions on the authenticity of the output. Although publishers such as Elsevier have endorsed the use of AI in scientific writing subject to conditions (Alqahtani et al., 2023).

The use of generative AI within research raises as authorship and copyright as key intellectual property concerns. Although they seem to be ignored within the institutions and regulatory frameworks such as the EU’s AI Act. In contrast with the US which through court litigation has demonstrated the significance of copyright within AI governance (La Diega & Koolen, 2024). Litigation generally relates to whether the AI generated output needs to be protected. UNESCO (2023a) is pressing for states to regulate the use of generative AI whilst paying attention to copyright issues. Generative AI providers such as Gemini, Open AI, and Copilot state that the end users own the output, in contrast many universities view the AI generated output as not being the user’s own intellectual creation (La Diega & Koolen, 2024). This begs the question of should a policy exist, which would advocate copyright over these outputs? If so, how would it plan out in the education domain? Questions also arise to if creative choices of prompts justify ownership. The final deliberation is to consider whether individuals place themselves at risk of copyright infringement by utilising generative AI outputs. Universities have tried to overcome this through demanding students provide the prompts used within the appendix. However, ultimately this demand itself lays out the possibility to expose users of copyright infringement.

2.2.1 Other Miscellaneous Concerns

There are concerns that university teachers will be merely seen as facilitators rather than subject specialists, leading to the potential loss of faith in the higher education sector. Tutors face threats of being undervalued, these threats align with the financial pressure faced by universities (Chan & Tsi, 2023). Other challenges include the quality and accuracy of the data retrieved by AI tools such as ChatGPT, as text is generated at times without any intellectual depth or authentic sources undermining intellectual rigour (Popenci, 2023). AI fabrication is very real as cyber criminals are outpacing the capabilities of authentic detection software, raising opportunities for misuse. A further area of concern is that tech-giants are manipulating the educational pedagogy and curriculum contents for HEIs. This is further destabilised by the AI arms race between the US and China (Rudolph et al., 2023). The control of data by the big techs exacerbates uneven distribution of wealth adversely affecting the public good. This is further amplified though LLMs being commercialised, which will create competing interests between the educational aims and capital interests (Klein, 2023). Finally, training of AI tools gives rise to environmental pollution as cloud computing, which is an essential component of generative AI, requires high energy consumption.

Inevitably, suggesting policy changes pose challenges at international, national, and institutional levels. The EU is set on regulating AI robustly, whilst the US would prefer international collaboration efforts. In line with the fast-paced evolution of LLMs the EU Artificial Intelligence Act poses a regulatory framework requiring the developers of these models to be transparent and safeguard fundamental rights (Artificial Intelligence Act, 2024). At the highest level placing a ban on high-risk applications posing unacceptable risks. HEIs have severely been impacted with the introduction of these LLMs such as ChatGPT. Institutions have begun to publish guidelines to safeguard student and staff interests. Following these challenges facing institutions in the post era of generative AI, there is a strong case for institutions to focus on pedagogical innovation coupled with initiating and implementing regulation alongside encouraging innovation. There is a need for education policies to promote a dynamic curriculum which works to integrate ethical AI education. Such policies need to support students to adopt self-directed learning in order to prepare them for lifelong learning to combat the challenges of evolving technologies. Policies, firstly, need to encourage industry partnerships, promoting opportunities to increase student employability. This should include compliance with legal and ethical frameworks. Such strategies require attention to data privacy, copyright, and intellectual property obligations. Secondly, they need to prioritise inclusivity and diversity to ensure that education is accessible to all members of society. Thirdly, policies need to entertain assessments which allow skills development, they need to support efficient resource allocation. Finally, policies need to support collaboration with third parties and establish forums for dialogue exchange to allow such policies to be evaluated and subjected to scrutiny making them suitable to the expectations for all involved groups (Abulibdeh et al., 2024). The next sections will focus on the current developments of AI policy nationally and globally, finally propose a potential roadmap for AI education policy.

2.3 *The Development of AI Policy around the Globe*

According to the artificial intelligence index report (see Fig. 10.1), AI products and services are favourably perceived in the global south compared to the global north. Over 70% of respondents in countries like Indonesia, Thailand and Mexico feel that products and services using AI have more benefits than drawbacks while less than 40% of respondents in countries like Canada, France and the United States believe that to be true. Yet 65% of the institutions in the Commonwealth countries do not have sufficient policies on AI in teaching and learning (Commonwealth of Learning, 2024). Despite this divide in favourability perception between countries, there is growing consensus among all countries on the need to govern the design, deployment and use of AI products and services. Therefore, countries around the globe have taken various measures to regulate AI, depending on their socio economic, political, and cultural practices. As mentioned earlier the EU has introduced the Artificial Intelligence Act, which reflects the EU's right centred approach (Mukerjee et al., 2023). In contrast the US is more lenient in adopting regulation at the federal

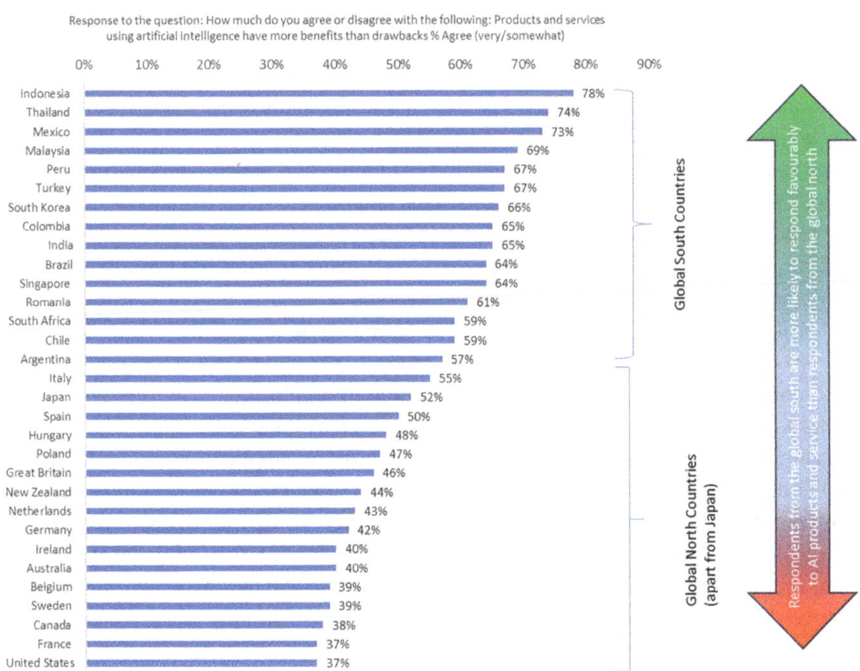

Fig. 10.1 Favourability perception of AI by country, adapted from Ipsos (2023)

level. It is heavily relying upon guidance from various industrial sectors, combined with state level regulatory frameworks (Larsen, 2023). Similarly, India has preferred to protect the current market dynamics, signalling the need to regulate if required in the future (Singh, 2023). The UK has protected public safety concerns through its future based policy commitments such as the AI governance framework, which will set out the government's expectations the ethical use of AI (UK Government, 2023). Whilst, Australia's voluntary AI Action Plan has committed to regulate AI's ability to provide deep fakes (Australian Government, 2023).

In response to the advancing implications of generative AI within education and research. UNESCO has released a guidance document generative AI in education and research (UNESCO, 2023a). The guidance is based on the UNESSCO's 2021 Recommendation on the Ethics of Artificial Intelligence, placing human rights, equity, and inclusion as key foundational principles for when using and developing AI tools (UNESCO, 2023b). The guidance recommends eight measures to use when initiating policy frameworks for deploying generative AI within education (UNESCO, 2023a). The first measure promotes inclusivity through adopting vigorous validation criteria when applying AI tools to ensure that such systems do not harbour bias and discrimination and support cultural and linguistic diversity. This is intended to make education and research more accessible and relevant on a global context. The second measure intends to protect human autonomy, encouraging human social interactions and accountability to promote human centric educational settings. This is more relevant when AI applications are relied upon to consider high stake decisions involving ethical concerns within educational settings. The third measure requires ethical validations before generative AI applications are adopted within the institutions. This requires steps for auditing to ensure inappropriately generated materials such as deep fakes are scrutinised, allowing the safeguarding of the wellbeing of staff and students. The fourth measure requires academic institutions to develop curricula, which supports students and staff understand the ethics related to the use of the AI applications. This should be conducted alongside providing regular training to staff and students to adapt to new technologies.

The fifth measure specifically requires staff training, highlighting their current competencies and areas for professional growth. This is to ensure that their rights are protected as they act as facilitators between students and the AI applications. The sixth measure requires educators to promote the generative AI responses. This is to encourage minority opinions and provide an exploratory environment which fosters learning though observations and empirical data collection. The seventh measure requires data collected through generative AI applications to be tested for its validity. The evidence collected from the same then should be tested for its environmental, social and ethical impacts. Following such outcomes, generative AI may be utilised in education when processing large scale data to further research. The final measure encourages measuring the long-term implications of the use of generative AI tools within education. This would encourage educators, AI providers, students, and researchers to adjust the curriculum and assessment methodologies.

2.4 Current AI Policy Formation in Education

A comprehensible policy framework is required nationally and internationally to harbour AI's growing potential within academia. This should examine global and national perspectives where institutions consider stakeholder needs to develop policies which are flexible, yet progressive. A global survey of over 10,000 university students from the UK, US, Canada, UAE, Singapore, Saudia Arabia, New Zealand and Australia highlighted that universities have been slow in adapting to the AI tools available within the market and that they need to rapidly integrate such tools within the teaching and learning framework (Bristow, 2024). On a national level, in 2024, the Higher Education Policy Institute's (HEPI) interviewed 1250 UK undergraduate students and found varying practices among students. 53% of the students admitted their use of generative AI tools, such as ChatGPT, is confined to explaining concepts, summarising research and editing text for them (Freeman, 2024a, p.4) while only 5% of students admitted to using AI tools to produce assessments without any editing done by them (Freeman, 2024b). At the same time, the HEPI study acknowledged that there is a 'digital divide', which may impact students access to AI. They reported that 58% of students from higher socio-economic backgrounds have admitted to the use of AI tools in comparison to 51% of students from lower socio-economic backgrounds (Freeman, 2024a). Therefore, there is a clear need to offer some guidance and training to students on the responsible use of AI. This training may include prompt engineering, identifying hallucinations, biases, and information inaccuracies. To combat the digital divide between students, institutions are encouraged to fund the provision of digital tools for students and teach them how to use such tools to expand on their digital literacy skills (Freeman, 2024a, p.1; Haire, 2024). This is imperative as only 9% of students interviewed in the HEPI study agreed that their institution was providing them with effective AI tools, while only 22% of students were content with the support provided by their institution on the use of AI tools (Freeman, 2024a, p.1). Consequently, HEPI recommended that institutions need to develop AI policies which support the use of generative AI within assessments and learning (Freeman, 2024a). Furthermore, on a national level, regulators such as the Department for Education need to publish routine guidance to assist institutions to identify assessment challenges within teaching and learning (Freeman, 2024a, p.15). This is due to the ever-increasingly sophistication of AI tools being introduced into the learning technology market with an ever-expanding capability to undertake various tasks, which threatens the perception of academic integrity within traditional forms of assessment. In conclusion, policies or guidance around AI should help bridge the AI divide between various student groups, help students develop responsible attitude towards AI and help key stakeholders keep abreast with latest developments within the AI-powered learning technology market.

Advocating for ethical and responsible use of AI technologies in education is key to establishing a future in which these emergent applications can be used positively from the start. The Russell Group principles on the use of generative AI tools in education provide universities with guidance to adopt assessment and teaching to anchor the ethical usage of AI generative application (Russell Group, 2023). In 2024 within a series of collaborative roundtable discussions British universities namely the University of Greenwich, University of Glasgow, Imperial College London, and Queen University Belfast shared their institutional frameworks to address the challenges posed by generative AI and assessments (Curnock & Brennan, 2024). The discussions highlighted that the leadership within the institutions can support students by ensuring that the assessments promote inclusivity and utilise the available resources. Glasgow University has devised a reflexive outlook to the adoption of AI. Staff are required to complete a survey which then correlates to a dashboard setting out how the staff member's assessments align with the University's assessment framework. This provides staff with the opportunity to contact resources and people who can assist to ensure that assessments are clear on outcomes and results from the beginning (Curnock & Brennan, 2024). The University of Greenwich has developed an AI Risk Measure Scale (ARMS) pointing out assessments which allow the misuse of generative AI, to be applied by the programme leaders against all assessments. This scale is designed to ensure that high scoring assessments on the scale are duly modified. This step is designed to ensure that generative AI is used and approached within assessments whilst conforming with ethical standards (Curnock & Brennan, 2024). Whilst the staff at Imperial College have begun to test their assessments via prompt engineering to test the rigour of the set assessments for taught module assessments. The results are shared widely amongst the teaching staff to ensure that assessments meet the requisite rigour if they are susceptible to generative AI misuse (Curnock & Brennan, 2024). In contrast Queen's University have imported augmented AI as part of their feedback provided to students. This approach to feedback has yielded positive results from by non-native speaking students (Curnock & Brennan, 2024).

The AI divide may be tackled through developing incentives for technology professionals to collaborate with universities, which would include teaching and contributing to the AI policies within the academic institutions. This type of integration can lead to professionals who are engaged within the industry to provide students and staff with insights on the most up to date developments in AI. Thereby, providing an enriching experience that bears relevance to the real world. It is imperative that policy makers, enterprises and educational leaders collaborate to produce innovative techniques to manage the challenges presented by AI. This multifaceted approach should aim to maintain competitiveness and promote growth within an era of expansive use of AI tools. There is also the need for institutional policies to set out strategies for dealing with uncertain information and the recognition of disinformation operations. This type of media literacy programme requires modules in curriculums to develop tools which allow students to understand AI's potential to spread disinformation and impact societal resilience (Shah & Soonawalla, 2024).

2.5 *A Forecast on the Role of Generative AI in the Higher Education Sector*

It is predicted that following the expanding use of AI tools, in particular LLMs such as generative AI applications, assessment and pedagogy in higher education will face many changes. This is inevitable as prospective employers will require graduates to possess proficient skills to utilise AI applications ethically (Colloton, 2024). The curricula will gradually include generative AI education. These changes will require universities to fund the integration of generative AI within the curricula, resulting in the greater use of technology driven assessments. The speed of uptake of such methodological changes to assessments will be influenced by the individual institution's priorities and personal AI transitional regulatory framework. It may be that institutions may revert back to face-to-face exams, whilst some may take the opportunities to set generative AI supported assessments. Those institutions which are more agile and pace their assessments methodologies in partnership with generative AI sooner than later will take greater benefit from industry partners who may want to train their workforce from suitable institutions. By utilising AI-centric educational paradigms, competent institutions may market themselves as being in-demand to cater towards the growing market of AI consumerism.

In terms of research, generative AI can be used to summarise previous works, assist with referencing and enhance writing styles. Specialised generative AI may be utilised to examine and verify the type of data analysis and methodological techniques. At the same time, it could be used by institutions to uphold academic integrity by supporting users to self-declare its usage and promote self-regulation. Generative AI may be able to assist with writing grant proposals, however transparency and auditability may be impacted as generative AI is exploited within the writing process. It needs to be used to enhance the quality of the application rather than simplify the procedure (Parrilla, 2023). Generative AI may also be used within the peer reviewing processes to improve transparency and validation of journal articles.

Whilst this is speculative it is predicted that generative AI will have a vast impact on education in the next 5–10 years. It may promote the use of AI assisted virtual environments with virtual tutors and AI translation fostering an inclusive environment. At the same time, it may assist in identifying gaps in learning and increase in efficiency through automated tasks and resource allocation. It will increase personalised content for students based on their individual needs.

3 Recommendations for a Roadmap for Developing an AI Education Policy

This section will outline a roadmap for devising an AI education policy for higher education institutions. The roadmap seeks to endeavour stakeholder participation to promote the use and adoption of AI technologies ethically within education.

Through this engaging process, active stakeholder participation can ensure that they are consulted to achieve an overall inclusive policy.

3.1 Establish a Steering Committee

The primary phase of this roadmap requires the planning and initiation of embedding AI policy within institutions via the establishment of a steering committee led by management. This committee would consist of IT experts, faculty members, students, and external stakeholders, working together to set a needs assessment to identify the gaps within the current policy. They would authorise data collection through online surveys, workshops and focus groups to determine the current challenges that a future AI policy should address.

3.2 Create a Stakeholder-Led Policy

Secondly, experts from within the steering committee would need to form a working group to draw up an AI policy. The policy would include legal and ethical considerations that would have been discussed within the steering committee meetings and throughout the stakeholder engagement. This policy document should then be revised following its input from all the relevant stakeholders following their timely feedback. Following this revision procedure, the policy would be formally adopted in readiness for its implementation.

3.3 Policy Content and Coverage

(a) Aims and goals.

The authors recommend that a policy of such nature should firstly have a clear communication strategy by way of a plan, which outlines the exact use of AI within the institution. Secondly, the policy should provide details on the guidelines used by the institution for the procurement of AI tools. The preparation and planning of such procedure should include AI impact assessments and details of market engagement. Finally, the policy should set out the criteria applied for selection and awards of such contracts, containing details of the expertise of the individuals involved in those processes. It should also identify the goals and the intended use of the AI tools particularly when used for research in higher education. The possible goals for use of AI within research would include (these are subjected to constant monitoring as AI tools continue to expand and refine their capacities):

1. Utilisation within the data collection stages such converting and interpreting data such as analysing voice calls or enabling individuals to express their opinions in the natural language. Summarise text and audio, to include converting emails and records into structured records.
2. Enhance research through retrieval of information.
3. Generate first drafts of documents.
4. Synthesise complex data to allow others to understand the text through producing simple summaries.
5. Allow greater accessibility of information through translation of different languages.

At the same time the policy should indicate the limitations of the usage of AI in research and that it should be avoided in the following areas:

1. During the making of fully automated decisions using the research involving any data which involves making significant decisions on areas such as an individual's health and safety, fundamental rights or the environment.
2. Generative AI must not be used on its own for high risks areas. This would include where there maybe any potential risks to breach of fundamental rights, the environment and health and safety legislation.
3. It should not be used in high explainability contexts as the inner workings of a generative AI based solution may be hard to explain.
4. As generative AI relies on a high volume of data, it should not be applied for research with limited data or in specialist areas using medical or legal terminologies. The risk is that such usage would lead to skewed results.
5. Generative AI should not be used in research where a rapid low latency result is needed as AI operates relatively slower to other computer systems.
6. Generative AI is not accurate due its tendency to produce hallucinations, therefore it should not be validated as the sole truth. Rather other measures should be operated to ensure accuracy.

(b) Skills.

The policy needs to ensure that there is an 'acquiring skills' section which outlines the skills required to utilise generative AI within the higher education settings. These skills would range from courses providing training on the ethical implications of AI, to more advanced technical courses looking at how to address the challenges of generative AI. This would include courses for researchers on the use of generative AI during the collection, visualisation and organisation of data. Secondly, providing advanced training on the technical aspects of utilising generative AI to researchers with interests in digital technologies. Thirdly, training should be provided on the operation of generative solutions to allow users to have the appropriate generative solutions designed for them to meet their usage criteria and achieve reliable results. Training would entail the understanding of the infrastructure of the operating mechanisms such as user prompts/user chatbox/cloud solutions.

(c) Data management

The policy will need to pay attention to data management strategies. Effective management of data is essential for implementing generative AI solutions (Government Digital Service, 2025). The policy should outline the processes and methodology that will be utilised to collect metrics from the AI usage. This will enable the institution to ensure safe compliant use and monitor its performance. The metrics will include user feedback from their interactions with the AI tool, allowing the institution to point out problematic behaviours of the tool (i.e. hallucinations and biases). This should allow better control of the generative AI tool improve the AI system. The policy will also need to ensure that the institution will take ownership for any risks and will bear responsibility for compliance. The data management process within the policy should include procedures to ensure that the use of personal data is not misleading or detrimental to individuals. Personal data is only used for clearly defined purposes, and it is not unjustifiably accumulated.

(d) Law

The policy will need to address the legal issues that arise from the use of AI within higher education institutions. It will need to make references to the appropriate legal teams to ensure that the institution is compliant with the data protection regulatory framework to include the UK General Data Protection Regulation (GDPR) and the Data Protection Act 2018. The policy will need to mention that the legal experts will need to devise contracts to apportion the intellectual property and liabilities following any research conducted via the AI tools. Finally, the policy will need to refer to the Equality Act 2010 requiring regular impact assessments to be conducted to eliminate potential biases from the use of the AI tools.

(e) Ethics

The policy will need to address ethical issues to maintain that the institution is utilising AI tools in a transparent, fair, accountable and non-discriminatory manner. To ensure transparency the policy will need to provide for internal mechanisms to monitor the design and practice of the generative AI tool. Procedures need to be in place to explain to citizens the way the AI tool operates and the factors it considers during its decision-making cycle. To ensure accountability the policy should require naming parties within the institution during the lifetime of the research project. Identification of parties involved with the generative AI during the research project cycle allows for answerability. Clear traceable reporting mechanisms which are easy to understand for all stakeholders need to be promoted within the policy from the innovation cycle from data collection until implementation. The policy needs to guarantee and provide evidence of audits for human involvement in automated processes, workflows and within the decision-making processes. As human involvement with AI processed ensures that the work supports human values and supports the public interests. The policy needs to provide insights on environmental considerations to include processes that minimise carbon footprint.

3.4 *Develop an Implementation Plan*

The adoption of the policy would lead to its implementation. There would be a need for the steering committee to develop an implementation plan to ensure that the policy is translated into practice. This would entail providing training and support mechanisms to staff, students and stakeholders to allow them to integrate AI within education.

3.5 *Monitoring and Evaluation*

The implementation plan would be followed through by a monitoring and evaluation team. The team would provide regular updates through drafting reports as insights on the impact of the policy on educational practices. Finally, further feedback would be collected to ensure that the policy could be revised to reflect any advancements in technology and the sector (Chan, 2024).

There are suggestions to develop a university data bank large language model, which would pool resources from the institutions to ensure that incidents of hallucinations; data security and equitable access to information are eliminated and documented (Stephenson, 2024). A viable roadmap needs to ensure that throughout the policy making and implementation processes key challenges mentioned earlier are addressed by institutions. Educational institutions need to firstly, ensure that teachers receive ongoing professional training for them to in return utilise AI platforms for student progress and lower their burden through optimising the use of automated administrative tasks. Tutors needs to be encouraged to utilise such platforms to share best practice and resources. This will free them to facilitate and mentor students. Teachers can also be supported through the provision of educational resources offering guidance on the capabilities of the large language models and opportunities for regular feedback to ensure that they can adjust for the secure ethical use of such applications. The roadmap needs to contain mechanisms which shift the focus from the sole dependence on AI for answers to a co-production of knowledge with the use of other academic resources. This will serve to ensure that AI is enhancing the learning experience rather than replacing the traditional modes of learning. Institutions need to also ensure that the AI model being deployed is up to date and providing the most current information, this would require regular review of the model's performance. It is acknowledged that the maintenance of large language models will be financially burdensome for some institutions. To mitigate for this risk, institutions may explore partnership with non-profit organisations to share resources and assist with the use of the large language models within education. Alternatively, institutions may invest in research to decrease the computational resources, model and data to lower costs.

Secondly, AI tools have the potential to transform tertiary education to make it more inclusive, the roadmap needs to ensure that such tools are developed to assist diverse learners including students with disabilities so that all students have equal opportunities to learn and engage. Mitigation strategies within the roadmap should entail regular monitoring measures to identify potential diverse groups which are potential to bias discrimination. Institutions need to ensure that they revise any algorithm biases to mitigate potential discrimination. This should also include bias-correction techniques such as post-processing methods. Transparency mechanisms should be introduced that allow the users to deliberate the model's output plus the data and any assumptions that were applied.

Thirdly, as AI applications become predominant within university learning practices, institutions need to prioritise ethical considerations which include adding concrete data protection safeguards, to ensure that data security is maintained. Proposals within the AI policy to mitigate for privacy and security include conducting regular audits of the privacy and security measures in operation; initiating response plans to respond to any breaches and being transparent with students about data collection and storage operations in use. Inevitably Investment in the IT infrastructure of institutions is needed, although this may be a challenge within itself as universities continue to be financially squeezed. The final unanswered question is whether the opportunities provided by AI outweigh the investments required to fulfil those opportunities.

4 Concluding Thoughts

AI applications in particular generative AI tools such as ChatGPT are transformative, as they offer learning opportunities such as providing instant feedback and efficiency where funding and human resource is limited. Yet they present ethical, practical and policy challenges for the HE sector. These include concerns related to cheating, plagiarism and an overreliance on AI tools limiting student motivation to develop their overall independent thinking and linguistic skills (Dwivedi et al., 2023). Aside from teaching and learning, academic integrity has also been impacted due to publications being published via ChatGPT acting as an author. This is problematic as AI is not responsible for the content, as a result devaluing the research output (Van Dis et al., 2023). The challenges noted by the use of generative AI within the higher education sector are present due to weak codes and policies around generative AI. A key challenge for the higher education sector is to educate students and educators on the limitations and benefits of such fast-paced applications. This includes developing policies which address trust, transparency and ethical issues if more reliance is to be encouraged of generative AI tools. Greater funding is required to ensure that institutions can test and apply these sophisticated applications for student learning. The limitation of this exploratory study is that it presents a snapshot of a given time period. It is anticipated that due to the advances and acceptance of usage of generative AI within education, policies will be constantly metamorphosing to meet the challenges of generative AI.

References

Abulibdeh, A., Zaidan, E., & Abulibeh, R. (2024). Navigating the confluence of artificial intelligence and education for sustainable development in the era of industry 4.0: Challenges, opportunities, and ethical dimensions. *Journal of Cleaner Production, 437*(140527), 1–15. https://doi.org/10.1016/j.jclepro.2023.140527

Alqahtani, T., Badreldin, A. H., Alrashed, M., Alshaya, A. I., Alghamdi, S. S., Saleh, B. K., Alowais, S. A., Alshaya, O. A., Rahman, R., Al Yami, M. S., & Albekairy, A. M. (2023). The emergent of artificial intelligence, natural processing, and large language models in higher education and research. *Research in Social and Administrative Pharmacy, 19*, 1236–1242. https://doi.org/10.1016/j.sapharm.2023.05.016

Artificial Intelligence Act (2024). Regulation (EU) 2024/1689 of the European Parliament and of the Council (Regulation EU 2024/1689), Official Journal of the European Union, L series. Available at: http://data.europa.eu/eli/reg/2024/1689/oj

Atadoga, A., Awonuga, K. F. C. U., Olu-lawal, K. A., & Usman, F. A. (2024). Harnessing data analytics for sustainable business growth in the US renewable energy sector. *Engineering Science and Technology Journal, 5*(2), 460–470. https://doi.org/10.51594/estjv.v512.806

Australian Government Department of Industry and Resources. (2023, June 18). Australia's artificial intelligence action plan. https://www.industry.gov.au/publications/australias-artificial-intelligence-action-plan. Last accessed 8/10/24.

Baker, R. S., & Hawn, A. (2022). Algorithmic bias in education. *International Journal of Artificial Intelligence in Education*, 1–41. https://doi.org/10.3542/osf.io/pbmvz

Bannister, P., Penalver, E. A., & Urbieta, A. S. (2024). International students and Generative artificial intelligence: A cross-cultural exploratory analysis of higher education academic integrity policy. *Journal of International Students, 14*(3), 149–170. ISSN:2166–3750.

Bozkurt, A., Xiao, J., Lambert, S., Pazurek, A., Crompton, H., Koseoglu, S., Farrow, R., Bond, M., Nerantzi, C., Honeychurch, S., Bali, M., Dron, J., Mir, K., Stewart, B., Costello, E., Mason, J., Stracke, C. M., Romero-Hall, E., Koutrospoulos, A., Toquero, C. M. S., et al. (2023). Speculative futures on ChatGPT and Generative artificial intelligence (AI): A collective reflection from the educational landscape. *Asian Journal of Distance Education, 18*(1), 53–130.

Bristow, I. (2024, April 15) Student voices on AI: Navigating expectations and opportunities. HEPI Blog. https://www.hepi.ac.uk/2024/04/15/student-voices-on-ai-navigating-expectations-and-opportunities/. Last accessed 8/10/24.

Burger, B., Kanbach, D., Kraus, S., Breier, M., & Corvello, V. (2023). On the use of AI-based tools like ChatGPT to support management research. *European Journal of Innovation Management, 26*(7), 233–241. https://doi.org/10.1108/EJIM-02-2023-0156

Butson, R., & Spronken-Smith, R. (2024). AI and its implications for research in higher education: A critical dialogue. *Higher Education Research and Development, 43*(3), 563–577. https://doi.org/10.1080/07294360.2023.2280200

Canterbury Christ University. (2023) (August 2023). Generative AI Guidance for Staff. chrome-extension://efaidnbmnnnibpcajpcglclefindmkaj. https://www.canterbury.ac.uk/learning-and-teaching-enhancement/resources-for-academics/Docs/Generative-Artificial-Intelligence-AI-Guidance-for-Staff.pdf. Last accessed 9/10/24.

Chan, C. K. Y. (2023). A comprehensive AI policy education framework for university teaching and learning. *International Journal of Educational Technology in Higher Education, 20*(1), 1–25. https://doi.org/10.1186/s41239-023-00408-3

Chan, C. K. Y. (2024). Chapter 5. Developing an AI in education policy. In C. K. Y. Chan & T. Colloton (Eds.), *Generative AI in higher education. The chat GPT effect* (pp. 127–165). Routledge.

Chan, K.Y., & Tsi, L.H. (2023). The AI revolution in education: Will AI replace or assist teachers in higher education? arXiv preprint arXiv:2305.01185.

Chaudhry, M. A., & Kazim, E. (2022). Artificial intelligence in education (AIEd): A high-level academic and industry note 2021. *AI and Ethics, 2*(1), 157–165.

Chinnasmy, P., Sathya, K. B. S., Jebamani, B. J. D., & Fowjiya, S. (2022). Deep learning. Algorithms, techniques, and applications—a systematic survey. In A. Kumar, Renuka, & S. Geetha (Eds.), *Deep learning research applications for natural language processing* (pp. 1–17). IGI Global. https://doi.org/10.4018/978-1-6684-6001-6.ch001

Christou, P. A. (2024). How we use artificial intelligence (AI) as a resource methodological and analysis tool in qualitative research? *The Qualitative Report, 29*(2), 560–576. https://doi.org/10.46743/2160-3715/2024.7046

Colloton, T. (2024). Chapter 7. The future of artificial intelligence in education. In C. K. Y. Chan & T. Colloton (Eds.), *Generative AI in higher education. The chat GPT effect* (pp. 238–254). Routledge.

Commonwealth of Learning (COL). (2024). Policy and Practice of Artificial Intelligence in teaching and learning at post-secondary educational institutions in the commonwealth. chrome-extension://efaidnbmnnnibpcajpcglclefindmkaj. https://oasis.col.org/server/api/core/bitstreams/f68f7f8b-100c-4d49-9675-f688b5ad9bf2/content. Last accessed 8/10/24.

Curnock, M., & Brennan, B. (2024, June 16). Using AI to think with students, not for them. HEPI Blog. https://www.hepi.ac.uk/2024/06/18/using-ai-to-think-with-students-not-for-them/. Last accessed 8/10/2024.

Duah, J. E., & McGivern, P. (2024). How generative artificial intelligence has blurred notions of authorial identity and academic norms in higher education, necessitating clear university usage policies. *The International Journal of Information and Learning Technology, 4*(2), 180–193. https://doi.org/10.1108/IJILT-11-2023-0213

Dwivedi, Y. K., Kshetri, N., Hughes, L., Slade, E. L., Jeyaraj, A., Kar, A. K., Baabdullah, A. M., Koohang, A., Raghavan, V., Ahuja, M., Albanna, H., Albashrawi, M. A., Al-Busaidi, A. S., Balakrishnan, J., Barlette, Y., Basu, S., Bose, I., Brooks, L., Buhailis, D., Carter, L., et al. (2023). Opinion paper: 'So what if ChatGPT wrote it' multidisciplinary perspectives on opportunities, challenges and implications of generative conversational AI for research practice and policy. *International Journal of Information Management, 71*, 102642. https://doi.org/10.1016/j.ijinfomgt.2023.102642

Freeman, J. (2024a, 1 February). Provide or punish? Students' views on generative AI in higher education. Higher Education Policy Institute. HEPI Blog. https://www.hepi.ac.uk/wp-content/uploads/2024/01/HEPI-Policy-Note-51.pdf. Last accessed 8/10 /24.

Freeman, J. (2024b, 1 February) new HEPI policy note finds more than half of students have generative AI for help on assessments—But only 5% likely to be using AI to cheat. HEPI Blog. https://www.hepi.ac.uk/2024/02/01/new-hepi-policy-note-finds-more-than-half-of-students-have-used-generative-ai-for-help-on-assessments-but-only-5-likely-to-be-using-ai-to-cheat/. Last accessed 8/10/24.

Gardner, J., O'Leary, M., & Yuan, L. (2021). Artificial intelligence in educational assessment: 'Breakthrough? Or buncombe and ballyhoo?'. *Journal of Computer Assisted Learning, 37*(5), 1207–1216. https://doi.org/10.1111/jcal.12577

Generative, A. I.. (2024, May 31). A technology that 'makes previously exclusive interactions available to all?'. (HEPI Blog 1). https://www.hepi.ac.uk/2024/05/31/generative-ai-a-technology-that-makes-previously-exclusive-interactions-available-to-all/. Accessed 8/10/24.

George Washington. (2023). (2023, April 2023). University Guidelines for using Generative Artificial Intelligence at the George Washington University. chrome-extension://efaidnbmnnnibpcajpcglclefindmkaj. https://provost.gwu.edu/sites/g/files/zaxdzs5926/files/2023-04/generative-artificial-intelligence-guidelines-april-2023.pdf. Last accessed 8/10/24.

Glendinning, I. (2022). Aligning academic quality and standards with academic integrity. In S. E. Eaton, J. G. Curtis, B. Stoesz, J. M. Kiata, C. Rundle, & J. Seeland (Eds.), *Contract cheating in higher education. Global perspectives on theory practice and policy* (pp. 199–218). Macmillan.

Government Digital Service (2025). Artificial Intelligence Playbook for the UK Government. Department for Science, Innovation and Technology. Available at:

https://www.gov.uk/government/publications/ai-playbook-for-the-uk-government/artificialintelligence-playbook-for-the-uk-government-html.

Haire, L. (2024, March 13). Generative AI in universities: what are educators thinking? HEPI Blog. https://www.hepi.ac.uk/2024/03/18/generative-ai-in-universities-what-are-educators-thinking/. Last accessed 8/10/24.

Hashem, R., Ali, N., Zein, F., Fidalgo, P., & Abu Khurma, O. (2024). AI to the rescue. Exploring the potential of ChatGPT as a teacher ally for workload relief and burnout prevention. *Research and practice in technology enhanced learning,* 19, 2023. https://doi.org/10.58459/rptel.2024.19023.

Hijazi, S. (2024, May 1). Universities: Guardians of ethical AI? HEPI Blog. https://www.hepi.ac.uk/2024/05/01/universitiesguardians-of-ethical-ai/. Accessed 8/10/2024.

Ibeh, C. V., Elufioye, O. A., Olorunsogo, T., Asuzu, O. F., Nduubuisi, N. L., & Daraojimba, A. I. (2024). Data analytics in healthcare: A review of patient-centric approaches and healthcare delivery. *World Journal of Advanced Research and Reviews, 21*(02), 1750–1760. https://doi.org/10.30574/wjarr.2024.21.2.0246

Ipsos. (2023). Global views on A.I. 2023: How people across the world feel about artificial intelligence and expect it will impact their life. IPSOS. Available at: https://www.ipsos.com/sites/default/files/ct/news/documents/2023-07/Ipsos%20Global%20AI%202023%20Report-WEB_0.pdf. Last accessed 8/10/24.

Kasneci, E., Sessler, K., Kuchemann, S., Bannert, M., Dementieva, D., Fischer, F., Gasser, U., Groh, G., Gunnemann, S., Hullermeier, E., Krusche, S., Kutyniok, G., Michaeli, T., Nerdel, C., Pfeffer, J., Poquet, O., Sailer, M., Schmidt, A., Seidel, T., Stadler, M., Weller, J., Kuhn, J., & Kasneci, G. (2023). ChatGPT for good? On opportunities and challenges of large language models for education. *Learning and Individual Differences, 103*(102274), 1–9. https://doi.org/10.1016/j.lindif.2023.102274

Klein, E. (2023). Opinion: The unsettling lesson of the OpenAI mess. *The New York Times*. https://www.nytimes.com/2023/11/22/opinion/openai-samaltman. html?searchResultPosition=4, Accessed 8/10/24.

Kooli, C. (2023). Chatbots in education and research: A critical examination of ethical implications and solutions. *Sustainability, 15*(7), 5614. https://doi.org/10.3390/su15075614

La Diega, G. N., & Koolen, C. (2024). Generative AI, education, and copyright law: An empirical study of policy making in UK universities. *European Intellectual Property Review, 46*(6), 346–360.

Larsen, B. (2023). The geopolitics of AI and the rise of digital sovereignty. The Bookings. Institute. https://www.brookings.edu/articles/the-geopolitics-of-ai-and-the-rise-of-digital-sovereignty/. Last accessed 8/10/24.

Liang, W., Yusekgonul, M., Mao, Y., Wu, E., & Zou, J. (2023) GPT detectors are biased against non-native English writers. *arXiv*. https://doi.org/10.48550/arXiv.2304.02819/.

Mollick, E., & Euchner, J. (2023). The transformative potential of generative AI. *Research-Technology Management, 66*(4), 11–16.

Mukerjee, S., Chee, F. Y., & Coulter, M. (2023, April 28). EU Proposes new copyright rules for generative AI. https://www.reuters.com/technology/eu-lawmakers-committee-reaches-deal-artificial-intelligence-act-2023-04-27/. Last accessed 8/10/24.

Noorbehbahani, F., Mohammadi, A., & Aminazadeh, M. (2022). A systematic review of research on cheating in online exams from 202-2021. *Education and Information Technologies, 27*(6), 8413–8460. https://doi.org/10.1007/s10639-022-10927-7

Nwokwdiegwo, Z. Q. S., Ugwuanyi, E. D., Dad, M. A., Majemite, M. T., & Obaigbena, A. (2024). AI-driven waste management systems: A comparative review of innovations in the USA and Africa. *Engineering Science & Technology Journal, 5*(2), 507–516.

Onesi-Ozigagun, O., Ololade, Y. J., Eyo-Udo, L. N., & Ogundipe, O. D. (2024). Revolutionizing education through AI: A comprehensive review of enhancing learning experiences. *International Journal of Applied Research in Social Sciences, 6*(4), 589–607. https://doi.org/10.51594/ijarss.v6i4.1011

Open AI (2023, January 31) New AI classifier for indicating AI written text. http://openai.com/blog/new-ai-classifier-for-indicating-ai-written-text. Last accessed 8/10/2024.

Parrilla, J. M. (2023 October 13). ChatGPT use shows that the grant application system is broken. Nature https://www.nature.com/articles/d41586-023-03238-5. *DOI:* https://doi.org/10.1038/d41586-023-03238-5 Last accessed 8/10/24, 443, 623.

Patton, C. V., Sawicki, D. S., & Clark, J. J. (2017). *Basic methods of policy analysis*. Routledge.

Perkins, M., & Roe, J. (2023). Decoding academic integrity policies: A corpus linguistics investigation of AI and other technological threats. A Preprint. 13 July 2023. British University Vietnam, Hung Yen Vietnam. James Cook University Singapore, Singapore.

Peters, M. A., Jackson, L., Papastephanou, Jandric, P., Lazaroiu, G., Evers, C. W., Cope, B., Kalantzis, M., Araya, D., Tesar, M., Mika, C., Chen, L., Wang, C., Sturm, S., Rider, S., & Fuller, S. (2024). AI and the future of humanity: ChatGPT-4, philosophy and education-critical responses. *Educational Philosophy and Theory, 56*(9), 828–862. https://doi.org/10.1080/00131857.2023.2213437

Popenci, S. (2023). The critique of AI as a foundation for judicious use in higher education. *Journal of Applied Learning and Teaching, 6*(2), 378–384. https://doi.org/10.37074/jalt.2023.6.2.4

Rudolph, J., Ismail, M. F., & Popenici, S. (2024). Higher education's generative artificial intelligence paradox: The meaning of Chatbot mania. *Journal of University Teaching and Learning Practice, 21*(6), 1–35. https://doi.org/10.53761/54fs5e77

Rudolph, J., Tan, S., & Tan, S. (2023). ChatGPT: Bullshit spewer or the end of traditional assessments in higher education? *Journal of Applied Learning and Teaching, 6*(1), 342–363. https://doi.org/10.37074/jalt.2023.6.1.9

Russell Group. (2023). The Russel Group Principles on the use of generative AI tools in education. https://russellgroup.ac.uk/news/new-principles-on-use-of-ai-in-education/. Last accessed 8/10/24.

Shah, S., & Soonawalla, A. (2024, January 12). The role for higher education in combatting AI misinformation. HEPI Blog. https://www.hepi.ac.uk/2024/01/12/the-role-for-higher-education-in-combatting-ai-misinformation/. Last accessed 8/10/24.

Singh, M. (2023, April 5). India opts against AI regulation. Tech crunch. https://techcrunch.com/2023/04/05/India-opts-against-ai-regulation/. Last accessed 20/8/24.

Stephenson, R. (2024, June 24). From knee-jerk reactions to saving the sector to opening the knowledge estate- discussions on generative AI. HEPI blog. https://www.hepi.ac.uk/2024/06/25/from-knee-jerk-reactions-to-saving-the-sector-to-opening-the-knowledge-estate-discussions-on-generative-ai/. Last accessed 8/10/2024.

Tan, S. (2023). Harnessing artificial intelligence for innovation in education. In *Learning intelligence: Innovative and digital transformative learning strategies: Cultural and social engineering perspectives* (pp. 335–363). Springer Nature.

Tozzi, A. E., & Cinelli, G. (2021). Informed consent and artificial intelligence applied to RCT and Covid-19. *BioLaw*, 97–108.

UCP. (2024). The university of reading classification policy. chrome-extension://efaidnbmnnnibpcajpcglclefindmkaj. https://www.reading.ac.uk/imps/-/media/project/functions/information-management-and-policy-services/documents/classification-policy-v11.pdf?la=en&hash=ED99CE3050345CE835BDF1E0CF0FDAF6. Last accessed 7/10/24.

UNESCO. (2021). AI and education. Guidance for policy-makers. UNESCO. http://unesdoc.unesco.org/ark:/48223/pf0000376709. Last accessed 8/10/24.

UNESCO. (2023a). Guidance for generative AI in education and research. UNESCO. October 13 2023. https://unesdoc.unesco.org/ark:/48223/pf0000385877?posInSet=2&queryId=296bf094-b045-4944-b82a-aef4154d1325. Last accessed 8/10/24.

UNESCO (2023b). UNESCO's recommendation on the ethics of artificial intelligence. Key Facts. https://www.unesco.org/en/articles/unescos-recommendation-ethics-artificial-intelligence-key-facts#:~:text=The%20Recommendation%20establishes%20a%20set,the%20rule%20of%20law%20%20online. Last accessed 8/10/24.

University of Reading. (2024). Data protection and AI. https://www.reading.ac.uk/imps/data-protection/data-protection-and-ai. Last accessed 7/10/24.

Usman, F. O., Eyo-Udo, N. L., Etukudoh, E. A., Odonkor, B., Ibeh, C. V., & Adegbola, A. (2024). A critical review of AI driven strategies for entrepreneurial success. *International Journal of Management of Entrepreneurship Research, 6*(1), 200–215. https://doi.org/10.51594/ljmer.v6i1.748

Van Dis, E. A. M., Bollen, J., Zuidema, W., Van Rooji, R., & Bockting, C. (2023). ChatGPT five priorities for research. *Nature, 614*, 214–226. https://doi.org/10.1038/d41586-23-00288-7

Xiao, P., Chen, Y., & Bao, W. (2023). Waiting, banning and embracing: An empirical analysis of adopting policies for generative AI in higher education. *ArXiv, 10.48550/arXiv.2305.18617.*

Chapter 11
The Horizon of AI in Academic Research: The Role of Human Factors to Optimise Performance

Joana Eva Dodoo and Griffiths Twum Appeaning

Abstract Artificial Intelligence (AI) presents a new paradigm to academic research and has received wide attention due to its potential to influence key aspects of education and scholarly works such as instructional design, technological development, and teaching methodology. Questions have been raised about how AI can be infused into research without compromising quality in learning and instruction. In particular, how AI can be designed to complement human factors. The nexus between AI and human factors in the context of academic research remains relatively less explored. Yet, a human factors perspective on the use of AI applications in academic research is critical to optimise performance of both the system and user. To address this gap in knowledge, a case study was undertaken to help shed more understanding on the following dimensions: (a) the role of the researcher in the application of AI enabled research; (b) Challenges with the use of AI for academic research; and (c) ways in which human factors contribute to addressing challenges with AI in academic research. The study identified prompt engineering, critical thinking as key enablers for optimizing academic research using AI. In addition, challenges such as compatibility of AI and human intelligence, environment and technology were found. It is recommended that designs of AI systems for academic research should be anchored on both humanness and efficiency.

Keywords Artificial intelligence · Human factors · Education · Research

J. E. Dodoo (✉) · G. T. Appeaning
University of Cape Coast, Cape Coast, Ghana
e-mail: jedodoo@ucc.edu.gh

X. Zhou, H. Al-Samarraie (eds.), *Institutional guide to using AI for research*, Advances in Artificial Intelligence in Education 2,
https://doi.org/10.1007/978-3-031-94809-1_11

1 Introduction

The current workplace is in high demand of a workforce who are equipped with the skills and competencies to navigate the Artificial Intelligence (AI)-infused world of work (Sanusi et al., 2022). Towards this end, education has a critical role to play in ensuring an AI-ready workforce (Pedro et al., 2019). The integration of AI technologies into research and education has revolutionalised academic work and has the potential to alter human-systems interaction. AI constitutes a new paradigm to academic research that is as resourceful as it is challenging. Resourceful because it provides an unprecedented increase in information for research. It is however challenging in the sense that, its use raises questions on how AI could be infused into research and still maintain quality in learning and instruction. In particular, how AI can be designed to complement human factors. For example, the introduction of ChatGPT and DALL-E into research have raised concerns from higher education institutions about the speed and scope with which AI is changing the teaching and learning environment (Bond et al., 2024). The use of AI platforms has thus intensified the debate on readiness, ethics, trust, impact and value addition of AI within the research and training space. Moreover, system controls with limited understanding by humans can potentially lead to negative consequences to the user (Howard, 2019).

Research has mainly looked at the issue from the perspective of developing advanced AI technologies. The development of advanced AI systems for education, and specifically academic research has gained much traction due to its potential to influence key aspects of education and scholarly works such as instructional design, technological development, teaching methodology and academic research (Ouyang & Jiao, 2021). It has been observed in the literature that, AI enhances research productivity through its capacity to introduce objectivity and repeatability into the research process (Burger et al., 2023).

Several domains of AI have been applied in education, particularly knowledge-based AI, where the potential of big data to process huge volumes of variety of data in a meaningful way has been exploited for education (Pedro et al., 2019) and academic research. For example, educational data mining and learning analytics have been cited as two specific forms where big data has supported education (Pedro et al., 2019). Further, AI systems have been engaged in several educational practices such as intelligent tutoring, teaching robots, chatbots, learning analytics dashboards, adaptive learning systems, human-computer interactions, automated assessment of digitized artifacts among others (Chiu et al., 2023; Ouyang & Jiao, 2021). The use of algorithmic applications in education has promoted personalized learning among students, provided automated assessment systems for teachers, and facial recognition systems for observing learner behaviour (Akgun & Greenhow, 2022). AI-aided research has the potential to transform knowledge in tremendous ways.

These benefits notwithstanding, a number of drawbacks have been highlighted. It should have ordinarily been the case that AI algorithms should rather enhance human intelligence rather than diminishing them with machines (Li & Gu, 2023). However, there is strong evidence to the contrary. Research has observed that the integration of AI into research generates overreliance on automated processes which may compromise the quality of research output (Limongi, 2024). For example, the risk of loss of human autonomy, privacy, security and data protection, and potential job displacement have been highlighted as major concerns with the integration of AI into education (Geetha, 2023; Jain, 2023; Ozmen et al., 2023; Pedro et al., 2019). In a related vein, Abulibdeh et al. (2024) raised ethical concerns in relation to AI-generated content and its negative impact on critical thinking, teacher and student competencies, and potential biases. Furthermore, AI algorithms have been criticized as magnifying societal inequity and perceived as biased against certain races (Kusner & Loftus, 2020; Ozmen et al., 2023). Aside from the benefits of reduced writing time using AI, it necessitates extensive fact-checking due to inaccuracies in citations and potential risks of plagiarism (Kacena et al., 2024). It further widens the digital divide between researchers who are proficient with AI tools and those who are not, which potentially leads to inequalities in research opportunities and outcomes (Besiroglu et al., 2024).

The above highlights certain perceived risks and challenges imposed by the AI-human interaction. This calls for a better understanding about AI and its effect on students. Based on this, there is the urgent need to understand how human factors could empower students to fully use AI systems optimally. However, research on AI integration in education has rarely explored the role of human factors or given it the needed focus it deserves in scholarship. Human factors is a discipline that seek to understand the human-system interaction, by designing systems that are compatible with the needs, abilities and limitations of humans in order to optimise the well-being and the overall system performance (Karwowski & Zhang, 2021). Thus, designing AI systems that are humane can potentially enhance its application in education (Li & Gu, 2023) and academic research. It will contribute to addressing concerns that have been raised in relation to safety and wellbeing of humans and the need to mitigate the unfavourable effects of AI on users (Howard, 2019). Also, it will lead to a balanced, responsible and healthy use of digital technology (Chiu et al., 2024), and understanding of human abilities, their limitations, and implications on the design of these technologies.

To address this gap in knowledge, a case study research approach was used to understand users' (postgraduate students, researchers and lecturers) insight on how human factors could optimise the use of AI in academic research. The following specific objectives guided the study: (a) the role of the researcher in the application of AI-enabled research; (b) challenges with the use of AI for academic research; and (c) how human factors contribute to addressing challenges with AI in academic research. The remainder of the chapter is organized as follows: Literature review, methodology, a case study on perspectives from study participants, and concludes with discussion and implication, conclusion and recommendations for future research direction.

2 Literature Review

2.1 *The Role of the Researcher in the Application of AI Enabled Research*

2.1.1 Human-System Interaction

Understanding the AI language for effective communication is a fundamental requirement for postgraduate students and teachers. It enhances research efficiency through automation of repetitive tasks and its robust data analysis capabilities (Burger et al., 2023). To fully leverage on the strengths of AI, its versatility, and capacity for collaboration, it is imperative that researchers acquire essential skills, such as the ability to craft precise AI prompts (Prompt engineering), engage in critical thinking, and transparently disclose the use of AI in their scholarly work (Lin, 2023). To achieve clear communication with AI tools, effective human-system interaction is essential.

Human-system interaction (HSI) within the context of AI concerns the design, development and research on effective interactions between humans and intelligent systems (Wickramasinghe et al., 2020). HSI has implication on the performance of humans as well as the efficiency of the systems. Owing to this, information on human behaviour, their abilities and limitations, and other characteristics are applied into the design of technology for safe and productive use by humans. Human characteristics ranges from physical, cognitive, to emotion or motivation (Karwowski & Zhang, 2021). The combined effects of these characteristics influence the processing and interpretation of information between humans and systems. Factors that influence human-system interaction include usability, interpretability, and trust.

Usability is perceived as fundamental to how researchers can engage intuitively with AI tools (Wiemer et al., 2023). It fosters adoption and efficiency of using AI in research. Research has observed that AI-enabled software assists precision in citation screening better than non-AI tools (Xu et al., 2021). Moreover, AI tools support with multiple stages of systematic reviews, search development, screening, data extraction and summarization (Fabiano et al., 2024) which improves researchers' efficiency significantly. Interpretability is essential for understanding AI systems' output which by implication can potentially increase acceptance and generate trust (Carter-Browne et al., 2021). Trust in AI systems is generated through validation, continuous improvement, and transparency (Hou et al., 2021). Although the above presents principles that undergird the human-systems interaction, its integration into academic research creates some challenges.

2.2 *Challenges with the Use of AI for Academic Research*

The human-systems interaction poses challenges for researchers and system efficiency. A number of challenges such as usability, ethicality, value addition to academic discourses (Bond et al., 2024) have been highlighted in the literature.

Moreover, research has observed that, the ability of the AI system to optimise its own skills require a functioning IT system that understand both the human's physical properties and cognitive processes of relevant tasks in order to work autonomously. Owing to this, it is therefore, critical that the IT systems are comprehensively intelligent and able to communicate with other devices and the human being effectively (Wisskirchen et al., 2017). Further, environmental requirements have been highlighted in the literature as central to the successful integration of AI into academic research (Karwowski & Zhang, 2021). The above highlights the critical roles of both environmental and technological requirements in the conversation on AI integration and academic research. To understand the role of human factors in the integration of AI into academic research, this study, will place emphasis on environmental and technological requirements that pose as challenge.

2.2.1 Environmental Requirements (Physical and Digital)

Environmental requirements concern the conditions in and around the system that could potentially impact on the human's ability to function as part of the system (Karwowski & Zhang, 2021). This can be considered from two main perspectives, namely the physical and digital environments.

Physical

On one hand, the physical environment of a research setting is critical in determining the usability of AI tools. For instance, availability of suitable hardware, workspace design, and the general infrastructure of the research facility are important for effective use of AI technologies (Karwowski & Zhang, 2021). For high-performance, AI tools often require computing resources to be housed in temperature-controlled environments with adequate power supply and network connectivity (Hou et al., 2021). If these physical requirements are not met, the performance and reliability of AI tools can be compromised, limiting their usability. Additionally, the layout and ergonomics of the physical workspace can influence how easily researchers can access and interact with AI tools, affecting productivity and comfort during research activities. Environmental sustainability is important due to the significant energy consumption and carbon footprint implication involved with AI training and use (Konya & Nematzadeh, 2024).

Digital

On the other hand, the digital environment covers the software, platforms, and online resources that researchers use to interact with AI tools. For example, user interface design, data management systems, and the integration of AI tools with other digital research infrastructure (Ozmen et al., 2023). A well-designed digital environment facilitates the smooth operation of AI tools, providing researchers with intuitive interfaces and seamless integration with other software commonly used in academic research. For example, AI tools that are embedded within a comprehensive research management platforms can streamline workflows by automating data analysis, managing research projects, and facilitating collaboration among researchers. Conversely, if the digital environment is fragmented or lacks compatibility with AI tools, it can create barriers to usability, leading to inefficiencies and errors in the research output.

The interplay between physical and digital environments play a crucial role in determining the effectiveness of AI tools in academic research. A cohesive integration of physical infrastructure with digital platforms is essential for AI tools to operate at their full potential. For example, a well-equipped research lab with robust digital infrastructure allows researchers to leverage AI for real-time data analysis, simulations, and modeling without delays or technical issues (Wiemer et al., 2023). Meanwhile advanced AI software running on inadequate hardware can potentially lead to system crashes, slow processing times and reduced effectiveness. Given the above, environmental requirements are considered as critical for smooth functioning of AI technologies.

Research institutions in high-to-low-income countries differ widely in infrastructure, resources, and technological capabilities. As a result, adaptability of AI tools to a variety of physical and digital environments in diverse academic research settings are important for its effectiveness (Xu et al., 2021). For example, AI tools designed with adaptability (such as those that can operate on different types of hardware or integrate with various software platforms) are more likely to be effective in different environments (Carter-Browne et al., 2021). Moreover, tailored user training and support can significantly enhance the usability and effectiveness of AI tools across different settings. Therefore, ensuring their successful deployment in both resource-rich and resource-constrained research environments can contribute to addressing the challenges.

2.2.2 Technological Requirements (Reliability, Maintainability and Logistic Support)

Reliability

A significant challenge in AI application is the reliability of the system (Blood et al., 2023). Reliability concerns the consistency and accuracy of AI models in generating expected research outcomes across different datasets. AI systems rely on large dataset, and their performance is linked to data quality and diversity (Wiemer et al., 2023). For example, If the data used to train AI models are biased or incomplete, the AI output will be unreliable or skewed, leading to a flawed research conclusion. Also, the usefulness of AI systems is limited in different research contexts, which can undermine the credibility of research findings and make it difficult for researchers to fully trust and accept AI tools.

Maintainability

Maintainability presents yet another critical challenge. It refers to the ease with which AI systems can be updated, modified, or corrected as research progresses. Academic research often comprises evolving questions and methodologies, which require AI systems to be adaptable (Hou et al., 2021). However, many AI models are complex and rigid, making it difficult for researchers to either understand or modify its underlying algorithms (Bergstra & Bengio, 2012; Sculley et al., 2015). This complexity leads to challenges with maintainability, especially when errors occur (Leveson, 2012). In addition, poor maintainability results in increased downtime, higher costs, and a greater need for specialized technical support, which can slow the research process and reduce productivity.

Logistic Support

Logistic support involves deploying, managing, and scaling up AI systems within academic research environments. AI systems require substantial computational resources, such as powerful hardware and access to large datasets, which can be a barrier for researchers, especially those with limited budgets (Ozmen et al., 2023). Additionally, ensuring that AI systems integrate into existing research workflows can present significant logistical challenge. Without adequate logistic support, researchers struggle to fully utilize AI tools, leading to inefficient workflows, increased operational costs, and inaccurate research outcomes. Addressing these technological challenges require a comprehensive approach that includes a consideration of human factors.

2.3 *Contribution of Human Factors Towards Addressing Challenges with AI in Academic Research*

Given the speed with which AI is impacting on the lives of humanity, an understanding of the role of human factors is essential. The conversation on human-centred AI has raised a number of questions such as: is AI human well-being oriented and responsible? Does it respect privacy? Does it incorporates human-centred design and evaluation frameworks, and respect human cognitive processes in its interaction with humans? (Ozmen et al., 2023). Other researchers suggest that the basic human factors that must be considered in the design of technology for safe and productive use by humans are cognitive, physical, and emotion or motivation (Karwowski & Zhang, 2021). These considerations are necessary principles for developing AI systems that complement human capabilities and decision-making in academic research.

Researchers have enumerated some principles that should underpin the design of human-centred AI systems. For example, the need to design AI-system that provides an unbiased decision-making and facilitate human-AI learning (Cronholm & Göbel, 2022). Moreover, it is recommended that an AI-adaptive learning system should enhance learner experiences as well as promote adoption (Kabudi et al., 2022). Steyvers and Kumar (2023) suggest that to understand the effect of different design choices on human-AI interaction, identifying conditions for complementarity is essential for assessing human mental models of AI. Thus, it was observed that a hybrid AI intelligence systems (AI and human involvement) that potentially optimize human input and understand humans as collaborators will help prevent certain mistakes and failures caused by AI when it is working alone (Kamar, 2016).

Further, research has observed that certain cognitive, behavioral, and operational weaknesses among human researchers also impact on the research process. For example, weaknesses such as confirmation bias, hindsight bias, and systematic misunderstanding of statistics (Bishop, 2020) could impact on systems efficiency and interpretation of outputs from AI. To address these issues, effective human-AI collaboration is key towards enhanced performance. Effective human-AI collaboration requires that AI systems are designed to complement human cognitive processes (Steyvers & Kumar, 2023). Most importantly, it is suggested that AI systems should be designed to model and understand human cognitive biases in order to improve the synergy between human judgment and machine efficiency (Gulati et al., 2023). Moreover, incorporating human-in-the-loop design and a usage architecture that integrates both AI and human decision-making, can potentially contribute to addressing the limitations associated with both humans and AI (Harfouche et al., 2023). Given the above, integrating human knowledge into AI design would ensure that it provides understandable explanations of their predictions.

3 Case Study: Method

Due to the limited research on human factors and AI integration in academic research from a developing country context (i.e., Ghana), a preliminary study was conducted to explore understanding of postgraduate students and other users on extent of use of AI tools in academic research. A case study research design which employed the use of interviews on questions based on the study's objectives was engaged. A snowball sampling strategy was used with the objective of capturing a broad range of insight on how human factors could optimize the use of AI in academic research. The focus was on postgraduate students/researchers who use AI for academic research. Data were collected through individual interviews from a sample of 10 participants (postgraduate students, lecturers and researchers). An in-depth interview guide underpinned by the study's objectives and validated by two experts in qualitative research was used for the data collection. On the average, interviews lasted between 15 and 30 mins. Before each interview, a statement of consent covering the objectives and expectations of the interview was read and permission was sought to record the interviews with a digital audio recorder. The audio files were transcribed verbatim and the transcripts compared with the audio recordings to ensure accuracy.

Manual coding was used to identify reoccurring responses and variations within the responses. This approach was adopted due to the fact that questions and number of participants were not many to handle. Authors read through the generated transcripts and discussed responses based on main themes of the study. Some key questions asked included: Is AI technologies important for academic research? If yes, in what way has AI supported academic research? What are the basic human requirements needed to understand the AI language in order to communicate effectively with the system? Are AI tools and systems compatible with human intelligence and abilities? Are there challenges in relation to using AI systems for academic research? If yes, what are the technological and environmental challenges hindering effective and efficient use of AI tools for academic research? How can AI systems and tools be designed to be compatible with human intelligence and abilities? among others. Thematic content analyses were conducted to explore key themes from the text. In addition, text mining method specifically the Latent Dirichlet Allocation (LDA topic modeling) was used to uncover latent structures within the text in order to reveal certain underlying semantic themes. Given the repetitive and consistent identification of key themes across different participants and disciplines, thematic saturation appears to have been reached. The data provides a robust foundation for the study, capturing the essential considerations for integrating AI into academic research through the lens of human factors. Study participants span across various academic disciplines ranging from postgraduate students (Accounting Education, Biomedical Science, Environmental Engineering, Information Technology, and Civil Engineering), researchers (Robotics, Renewable Energy, Mechanical Engineering, Physician Assistant), to lecturer (Information Technology/Computer Science) (see Fig. 11.1 for interviewee characteristics).

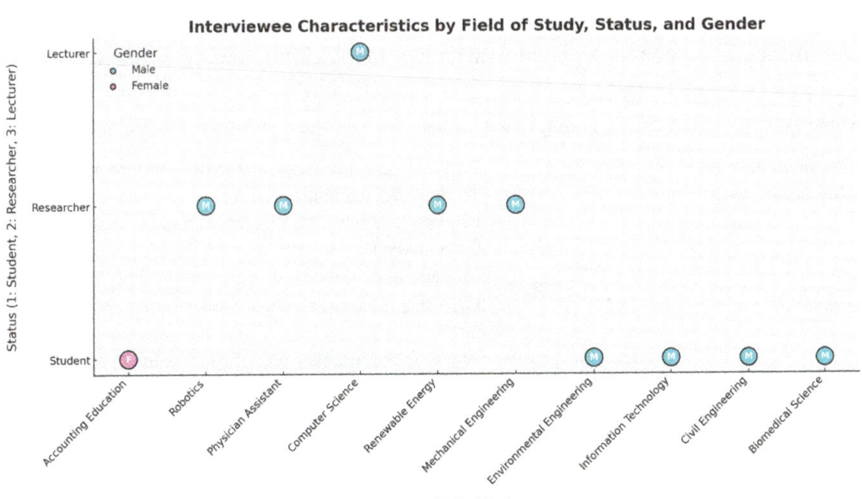

Fig. 11.1 Interviewee characteristics

4 Results and Discussion

4.1 *The Role of the Researcher in the Application of AI Enabled Research*

To understand participants insight on the role of the researcher in the application of AI enabled research, questions sought to explore the various ways AI has been utilised in their research work and benefits. From the analysis it was evident that participants have employed some AI tools such as ChatGPT, Zotero, Quilbot, among others for research purposes. In terms of usage, it came up strongly that AI has been primarily used as a tool to support concept generation and expansion of content, data analysis, literature reviews, and reference management. Further, participants noted that they have used AI to improve academic writing, and management of large dataset. Regarding the benefits, participants mainly expressed that, AI has supported them to clearly conceptualise ideas, breakdown concepts, and provides updated information on certain foundational concepts in research. Also, AI helps to summarise papers quickly, useful for proof-reading and correcting errors, and organise and interpret data faster. Further, AI has supported them to work more efficiently, allowing them to focus on core research tasks, particularly in literature search (key findings are summarised in Table 11.1). A participant remarked that

> *In my research, AI help me in data analysis and project management. I primarily use it to manage large environmental datasets, analyze trends in environmental impact assessments, and support decision-making processes for corporate social responsibility (CSR) projects* (Student, Male 7 (SM7)).

In addition, the word cloud LDA topic modeling analysis of the text applied in this study revealed an underlying co-occurring words/terms indicating the shared semantic themes in the text among participants in terms of use and benefit of AI for academic research. Word cloud depicts the word frequencies contained in a text or documents. The larger a depicted word/token in the cloud, the higher the frequency

Table 11.1 Summary of key findings on AI usage, benefits, and key human requirement

Application of AI into Academic Research	• Concept generation and expansion of content • To improve academic writing • Literature review and reference management • Data analysis • Manage large dataset
Benefits	• Facilitate understanding of foundational concepts of research • Works faster, quicker way for literature search, time saving
Key skills/attitudes	• Prompt engineering • Basic knowledge and understanding of research methods and analysis • Critical thinking

Fig. 11.2 Semantic themes on AI and the role of the researcher

of occurrence (Giannoulakis & Tsapatsoulis, 2021). The word cloud supports visualisation of relevant key themes that emerged from the data. It provides essential insights on themes that are linked with benefit of AI for academic research. Predominantly, a range of tokens such as tools, ChatGPT, research, managing, large, data, analysing, reviewing, literature, understand, concepts among others emerged. From the word cloud, it is clear that participants in this study mainly use AI tools such as ChatGPT for research, managing, processing large data, analysing, reviewing literature to understand concepts (see Fig. 11.2).

The study further sought for insights on basic human requirements such as skills and attitudes needed to understand the AI language in order to communicate effectively. The study categorised the findings into the following key themes:

4.1.1 Prompt Engineering

The findings showed that, prompt engineering is a critical skill needed to navigate AI tools for effective research outputs. It leads to effective communication with AI systems. This is because, AI can give broad, vague, and irrelevant outputs if one fails to query it correctly. Therefore, providing AI with precise and clear instructions has implication on the quality of data output. Prompt engineering skills and a good technical knowledge on how AI and machine learning works are relevant basic human requirement. It was clear from the findings that researchers' ability to give clear instructions and the right commands are key human requirements. A participant remarked that,

> *Prompt engineering is essential. It's about knowing how to ask the right questions or give the right commands. If you query an AI poorly, it will give you irrelevant answers. So, the critical skill is understanding how to communicate effectively with AI* (Student, Female 1(SF1)).

4.1.2 Basic Understanding of Research Methodologies

Another key human requirement was the need for researchers to have basic understanding of research methodologies, and data analysis tools and software. This is essential for constructing ideas clearly since AI cannot work effectively without proper inputs. It will also help them to interpret the validity of AI suggestions properly to suit their study's context. A participant explained that,

> *Familiarity with digital tools or data analysis programs is also important because it helps integrate AI into your workflow more smoothly* (Student, Male 8 (SM8)).

4.1.3 Critical Thinking and Understanding of the Limitations of AI

From the study's outcome, it is evident that AI as a supporting tool or system is not meant to replace the role of the human researcher, particularly critical thinking. AI lacks intuition and the ability to think creatively. It is not flexible as human thought and lacks the ability to apply ethical reasoning. Due to this, *critical thinking is crucial, researchers must be able to interpret AI's suggestions and decide whether they are valid* (Student, Male 10 (SM10)). The findings suggest the need for researchers to understand the place of AI, double-check its outputs, and refine responses to meet their needs. In addition, AI can help organize information, but it lacks the ability to fully understand or creatively generate new knowledge the way humans can. It is therefore imperative for researchers to understand the limitations of AI, and the need for human touch in order to interpret its outputs. Participants indicated that AI is still evolving and does not always give perfect answers. As a result, "*human oversight is essential in interpreting results and maintaining academic integrity*" (Researcher, Male 5 (RM5)).

4.2 Challenges Encountered with the Use of AI for Academic Research

In view of this objective, challenges hindering the effective use of AI for academic research were explored. The study categorized challenges based on the following key themes:

4.2.1 Compatibility

The study participants expressed ambivalent views in relation to AI tools and system compatibility with human intelligence and abilities. According to some, there is a clear lack of compatibility of AI tools and systems with human intelligence. Participants who held this viewpoint explained that AI systems lack intuition and emotional understanding, it lacks feelings and cultural awareness, and cannot make value judgements or interpret emotions. Therefore, these attributes of AI make it incompatible with human intelligence and abilities. However, some were of the view that, AI is compatible with human knowledge. A participant has this to say:

> *AI works based on algorithms created by humans, and its data set is derived from human knowledge and decisions. It's essentially a reflection of what people think, and thus, it's very compatible with human knowledge* (Lecturer, Male 4 (LM4)).

A notable exception to the arguments above was the findings that generally point to a complementary role of AI in relation to academic research. Most of the participants expressed that AI complements human knowledge, skills and abilities. According to this perspective, compatibility should be based more on a partnership between the AI and humans. This is because, humans have a role to play. Findings pointed out that, AI's strength is in processing data quickly and spotting patterns while humans provide intuition, creativity and ethical judgement, direction, and critical thinking to interpret AI simulations for informed decisions. Thus, AI complements human intelligence but does not replace it. A participant remarked that:

> *AI excels at processing large datasets and automating repetitive tasks, but human researchers provide the necessary context and interpretation* (Researcher, Male 6 (RM6)).

4.2.2 Environmental Challenges

Aside from the above, environmental challenges hindering effective and efficient use of AI tools for academic research were explored. Findings showed that environmental challenges relate mostly to digital simulation and modelling. It was observed that, AI provides unrelated language models that are generic which may require manual adjustment to suit context needs. It came up strongly in the results that AI lacks understanding of the local context specific issues, as a result, it generates data that are biased towards Western research styles. For example, a participant remarked that, *there are inherent biases in language and culture which is a challenge and do not reflect the realities of the local ecosystems* (Student, Male 7 (SM7)). Due to this, AI outputs do not align with local context, and prevailing environmental data or specific regional challenges. Meanwhile, AI's ability to interpret domain-specific language and contextual nuances are essential for research outcomes. This leads to a high probability of AI misinterpreting data. A participant remarked that:

> *As a Ghanaian, I may want to express something in a certain way, but AI may not fully understand due to its data set being skewed toward Western contexts (LM4).*

To address these challenges, it was recommended that designers consider creating AI that reads and summarizes large texts and offers insights specific to local contexts.

4.2.3 Technological Challenges

The findings showed that AI systems subscription fees and licenses, and supporting computing resources are expensive and non-affordable to certain groups of people. Due to this a number of individuals who are not able to afford payment for premium services opt for free versions which limits their research outputs as well as their experience with the system. Moreover, participants pointed out that, challenges relating to lack of high speed and stable internet to support advanced AI tools hindered effective integration of AI into academic research. A participant remarked:

> *Another challenge is access to advanced AI tools, as they can be costly, and some require high-speed internet, which is not always available in certain parts of the country* (PSM 8).

The findings further pointed that, AI is not reliable for performing nuanced tasks and interpreting complex research data and complex system interactions. AI tools lacks precision in academic context which requires advanced versions. A participant explained that: "*when I use AI to summarize articles, it sometimes misses important points or struggles with specialized engineering terms*" (PSM 9).

4.3 How Human Factors Contribute to Addressing Challenges with AI in Academic Research

To gain insights into the role of human factors and the integration of AI into academic research, questions on how AI tools and systems can be designed to be compatible with human intelligence and abilities were explored. The findings point out the need for human intelligence to direct AI as well as adjust output to align with local context. Since, AI lacks the intuition and the understanding of a subject that humans have. Also, it was recommended that an AI that is more intuitive, and can understand the nuances and represent different languages and cultures better will suffice. For example, a participant remarked that, "*I want it to be more intuitive about context, especially local practices and challenges*" (Researcher, Male 3 (RM3). Further, majority of the participants indicated the need to improve AI's context awareness and make it smarter and adaptable to specific research field. This will contribute to addressing concerns related to the generic nature of outputs from AI. Therefore, AI that is modelled to understand local context, languages and practices is critical for academic research performance. Moreover, the word cloud LDA topic modeling analysis of the text on this objective confirms the narratives above. It revealed co-occurring words depicting relevant key themes in the text or document. From the word cloud visualisation, it is clear that participants perspectives point to a range of tokens such as human, intelligence, think, data, creativity, process, knowledge, interpret among others. It further buttresses the superior role of human intelligence, and creativity in academic research.

5 Discussion and Implications

The study highlights that AI has been used primarily as a tool to support tasks such as managing and analysing vast amounts of data, facilitating literature reviews, data sorting, and pattern recognition, thereby saving researchers considerable time and effort. The results supports similar research findings that found that AI has been utilised in research activities such citation management, multiple stages of systematic reviews, screening, and data extraction (Fabiano et al., 2024; Xu et al., 2021). Moreover, the findings highlighted three major basic human skills and attitudes required for successful integration of AI into academic research. First, prompt engineering. Ineffective communication with AI was noted as the cause of the many inaccurate and general research output. The finding confirms the proposition on the need for researchers to acquire essential skills such as the ability to craft precise AI prompts for a better results (Lin, 2023). It calls for a deeper engagement with AI tools by researchers, providing contextual inputs and critically assessing AI outputs to bridge the gap between AI's general capabilities and specialized research needs. Second, a basic understanding of research methodologies and data analysis software by researchers is essential for interpreting the validity and suitability of AI outputs. This finding confirms the observation that students' lack of understanding of research methodologies and statistics impacts negatively on the interpretation of AI outputs (Bishop, 2020). Third, critical thinking was underscored as a crucial human requirement. It came up strongly that, AI lacks intuition, creativity, ethical judgment, and contextual understanding. Owing to these limitations, critical thinking from the human user is necessary for successful integration into academic research. According to Lin (2023), critical thinking is an essential skill for effective human-system interaction.

These notwithstanding, some major challenges related to compatibility, environment and technology were found with the integration of AI system into academic research. Regarding the lack of compatibility, the finding emphasises on the role of human touch, and the need for human-AI collaboration in academic research. This assertion supports existing research that emphasises that effective human-AI collaboration is essential for enhanced performance (Steyvers & Kumar, 2023). Moreover, environmental challenges were mostly found with unrelated language models and simulations that are skewed towards Western contexts. AI was blamed for lacking context understanding, and providing general and biased data which are skewed to Western context. These among other challenges found require effective human-system interaction. There is the need for AI systems that are human-centerd, and complements the academic research journey. This challenge was also highlighted in previous studies as an enabler of societal inequity (Kusner & Loftus, 2020; Ozmen et al., 2023). In terms of technological challenge, results point to reliability and inaccuracy of AI output which may impact negatively on academic integrity.

Additionally, the study's findings on the interplay of human factors and AI highlight two major human characteristics such as cognitive and emotion. Cognitive attributes such as creativity, logical reasoning, critical thinking, and ethical judgement were noted as human characteristics lacked by AI. It was noted in literature that work activities such as multitasking, computation, and memory can be best performed by AI however, humans excel in logical reasoning, language processing, creativity and emotion (Ozmen et al., 2023). Further, emotions, feelings and intuition were also found to be missing with AI. The lack of these human characteristics in AI had implication on the accuracy of its research outputs.

In order to optimise the use of AI and research performance of postgraduate students, the study proposes the following as areas of improvement; creating AI tools that reads and summarizes large texts and offers insights specific to local contexts. Also, AI that is modelled to understand local languages and practices. This will ensure that AI outputs are interpreted and utilized effectively within the nuanced frameworks of academic research. The study's finding further places emphasis on the fact that AI should not be viewed as a replacement of human characteristics but considered as a support tool. This can be achieved by training researchers to communicate effectively with AI in order to leverage on its computational power to maximise research outcomes. Postgraduate students must balance AI automation with critical human analysis to ensure that research processes remain efficient and accurate. Moreover, the high cost of financial and infrastructural resources restrict access to advanced AI tools. These have implication on the type of research output as well as widen the inequality gap. It proposes that affordable, scalable AI solutions, and improved infrastructure are critical enablers of broader AI adoption. Additionally, postgraduate students are encouraged to develop skills in crafting precise AI prompts, using AI for repetitive tasks such as data sorting and analysis, while focusing on their human strengths of creativity, critical thinking, and ethical judgment.

6 Conclusion and Limitation

AI's strength in data management underscores its potential to augment human capabilities in research. However, the reliance on AI for data-intensive tasks necessitates a robust understanding of AI systems by researchers to ensure research integrity and relevance. The complementary nature of AI and human intelligence highlights the importance of human factors alongside AI's proficiency. Training researchers to effectively interpret and leverage AI outputs is crucial for optimising academic research and performance. Moreover, enhancing AI's contextual understanding through domain-specific training and localized data integration can significantly improve its reliability and relevance in specialized academic disciplines. This requires a collaborative effort between AI developers and subject matter experts to model AI systems to meet specific academic needs. Further, it is recommended that the integration of structured oversight mechanisms and ethical guidelines into AI is essential for academic research. Training researchers on ethical use and protocols for AI supervision can enhance the reliability and ethical compliance of AI-assisted research.

However, the study is limited in the use of small sample size and insights from Ghana only, which may affect the generalization of the findings. Yet, since this is a preliminary study, it is recommended for future research to consider a bigger sample size as well as expand the scope of research area. Moreover, since the findings on human factors captured only cognitive and emotion characteristics, we call for further research to consider how AI could impact on the physical characteristics of researchers.

References

Abulibdeh, A., Zaidan, E., & Abulibdeh, R. (2024). Navigating the confluence of artificial intelligence and education for sustainable development in the era of industry 4.0: Challenges, opportunities, and ethical dimensions. *Journal of Cleaner Production*, 140527.

Akgun, S., & Greenhow, C. (2022). Artificial intelligence in education: Addressing ethical challenges in K-12 settings. *AI and Ethics, 2*(3), 431–440.

Besiroglu, T., Bergerson, S. A., Michael, A., Heim, L., Luo, X., & Thompson, N. (2024). The compute divide in machine learning: A threat to academic contribution and scrutiny? *arXiv preprint arXiv:2401.02452*.

Bergstra, J., & Bengio, Y. (2012). Random search for hyper-parameter optimization. *Journal of Machine Learning Research, 13*, 281–305.

Bishop, D. V. (2020). The psychology of experimental psychologists: Overcoming cognitive constraints to improve research: The 47th Sir Frederic Bartlett lecture. *Quarterly Journal of Experimental Psychology, 73*(1), 1–19.

Blood, J. C., Herbert, N. W., & Wayne, M. R. (2023). Reliability assurance for AI systems. Paper presented at the 2023 annual reliability and maintainability symposium (RAMS).

Bond, M., Khosravi, H., De Laat, M., Bergdahl, N., Negrea, V., Oxley, E., et al. (2024). A meta systematic review of artificial intelligence in higher education: A call for increased ethics, collaboration, and rigour. *International Journal of Educational Technology in Higher Education, 21*(1), 4.

Burger, B., Kanbach, D. K., Kraus, S., Breier, M., & Corvello, V. (2023). On the use of AI-based tools like ChatGPT to support management research. *European Journal of Innovation Management, 26*(7), 233–241.

Carter-Browne, B. M., Paletz, S., Campbell, S. G., Carraway, M. J., Vahlkamp, S. H., Schwartz, J., & O'Rourke, P. (2021). There is no "AI" in teams: A multidisciplinary framework for AIs to work in human teams. *Applied Research Laboratory for Intelligence and Security (ARLIS) Report June*.

Chiu, T. K., Ahmad, Z., Ismailov, M., & Sanusi, I. T. (2024). What are artificial intelligence literacy and competency? A comprehensive framework to support them. *Computers and Education Open, 6*, 100171.

Chiu, T. K., Xia, Q., Zhou, X., Chai, C. S., & Cheng, M. (2023). Systematic literature review on opportunities, challenges, and future research recommendations of artificial intelligence in education. *Computers and Education: Artificial Intelligence, 4*, 100118.

Cronholm, S., & Göbel, H. (2022). Design principles for human-centred AI.

Fabiano, N., Gupta, A., Bhambra, N., Luu, B., Wong, S., Maaz, M., et al. (2024). How to optimize the systematic review process using AI tools. *JCPP Advances, 4*(2), e12234.

Geetha, P. (2023). *Comprehensive overview of the opportunities and challenges in AI.* Paper presented at the 2023 international conference on sustainable computing and smart systems (ICSCSS).

Giannoulakis, S., & Tsapatsoulis, N. (2021). Topic identification via human interpretation of word clouds: The case of instagram hashtags. 17th IFIP international conference on artificial intelligence applications and innovations (AIAI), Jun 2021, Hersonissos, Crete, Greece. (pp. 283–294), https://doi.org/10.1007/978-3-030-79150-6_23.hal-0328766.

Gulati, A., Lozano, M. A., Lepri, B., & Oliver, N. (2023). Biased: Bringing irrationality into automated system design.

Harfouche, A., Quinio, B., & Bugiotti, F. (2023). Human-centric AI to mitigate AI biases: The advent of augmented intelligence. *Journal of Global Information Management (JGIM), 31*(5), 1–23.

Hou, M., Ho, G., & Dunwoody, D. (2021). IMPACTS: A trust model for human-autonomy teaming. *Human-Intelligent Systems Integration, 3*(2), 79–97.

Howard, J. (2019). Artificial intelligence: Implications for the future of work. *American Journal of Industrial Medicine, 62*(11), 917–926.

Jain, R. (2023). The impact of artificial intelligence on business: Opportunities and challenges. Available at SSRN 4407114.

Kabudi, T., Pappas, I. O., & Olsen, D. H. (2022). *Deriving design principles for AI-adaptive learning systems: Findings from interviews with experts.* Paper presented at the conference on e-business, e-services and e-society.

Kacena, M. A., Plotkin, L. I., & Fehrenbacher, J. C. (2024). The use of artificial intelligence in writing scientific review articles. *Current Osteoporosis Reports, 22*(1), 115–121.

Kamar, E. (2016). Directions in hybrid intelligence: Complementing AI systems with human intelligence. Paper presented at the IJCAI.

Karwowski, W., & Zhang, W. (2021). The discipline of human factors and ergonomics. *Handbook of human factors and ergonomics*. (pp. 1–37).

Konya, A., & Nematzadeh, P. (2024). Recent applications of AI to environmental disciplines: A review. *Science of the Total Environment, 906*, 167705.

Kusner, M. J., & Loftus, J. R. (2020). The long road to fairer algorithms. *Nature, 578*(7793), 34–36.

Leveson, N. (2012). *Engineering a safer world: Systems thinking applied to safety*. MIT Press.

Li, S., & Gu, X. (2023). A risk framework for human-centered artificial intelligence in education. *Educational Technology & Society, 26*(1), 187–202.

Limongi, R. (2024). The use of artificial intelligence in scientific research with integrity and ethics. *Future Studies Research Journal: Trends and Strategies, 16*(1), e845–e845.

Lin, Z. (2023). Why and how to embrace AI such as ChatGPT in your academic life. *Royal Society Open Science, 10*(8), 230658.

Ouyang, F., & Jiao, P. (2021). Artificial intelligence in education: The three paradigms. *Computers and Education: Artificial Intelligence, 2*, 100020.

Ozmen, G. O., Winslow, B., Andolina, S., Antona, M., Bodenschatz, A., Coursaris, C., et al. (2023). Six human-centered artificial intelligence grand challenges. *International Journal of Human–Computer Interaction, 39*(3), 391–437.

Pedro, F., Subosa, M., Rivas, A., & Valverde, P. (2019). Artificial intelligence in education: Challenges and opportunities for sustainable development.

Sanusi, I. T., Olaleye, S. A., Agbo, F. J., & Chiu, T. K. (2022). The role of learners' competencies in artificial intelligence education. *Computers and Education: Artificial Intelligence, 3*, 100098.

Sculley, D., et al. (2015). Hidden technical debt in machine learning systems. In *Advances in neural information processing systems*. (pp. 2503–2511).

Steyvers, M., & Kumar, A. (2023). Three challenges for AI-assisted decision-making. *Perspectives on Psychological Science, 17456916231181102.*

Wickramasinghe, C. S., Marino, D. L., Grandio, J., & Manic, M. (2020). *Trustworthy AI development guidelines for human system interaction.* Paper presented at the 2020 13th international conference on human system interaction (HSI).

Wiemer, H., Schneider, D., Lang, V., Conrad, F., Mälzer, M., Boos, E., et al. (2023). Need for UAI–anatomy of the paradigm of usable artificial intelligence for domain-specific AI applicability. *Multimodal Technologies and Interaction, 7*(3), 27.

Wisskirchen, G., Biacabe, B. T., Bormann, U., Muntz, A., Niehaus, G., Soler, G. J., & von Brauchitsch, B. (2017). Artificial intelligence and robotics and their impact on the workplace. *IBA Global Employment Institute, 11*(5), 49–67.

Xu, Y., Liu, X., Cao, X., Huang, C., Liu, E., Qian, S., et al. (2021). Artificial intelligence: A powerful paradigm for scientific research. *The Innovation, 2*(4).

GPSR Compliance
The European Union's (EU) General Product Safety Regulation (GPSR) is a set of rules that requires consumer products to be safe and our obligations to ensure this.

If you have any concerns about our products, you can contact us on

ProductSafety@springernature.com

In case Publisher is established outside the EU, the EU authorized representative is:

Springer Nature Customer Service Center GmbH
Europaplatz 3
69115 Heidelberg, Germany

www.ingramcontent.com/pod-product-compliance
Lightning Source LLC
Chambersburg PA
CBHW061300091225
36515CB00003B/93
* 9 7 8 3 0 3 1 9 4 8 0 8 4 *